# *Last Days in Eden*

# Last Days in Eden

ELSPETH HUXLEY & HUGO van LAWICK

THE AMARYLLIS PRESS
NEW YORK

The Amaryllis Press, Inc.
212 West 79th Street
New York, N.Y. 10024

This book was designed and produced by
The Rainbird Publishing Group Limited
40 Park Street, London W1Y 4DE

House Editor: Jasmine Taylor
Designer: Lee Griffiths
Production: Lorna Simmonds
Map: Eugene Fleury

Library of Congress Catalog Card Number
84–070583

ISBN 0–943276–02–0

Text filmset in Palatino by Wyvern Typesetting Limited, Bristol, England
Color origination by Gilchrist Brothers Limited, Leeds, England
Printed and bound by Brepols s.a., Turnhout, Belgium

*Half-title*: Vultures, the true scavengers, roost at night in the
relative safety of trees or in rock crevices.
*Title*: Cold early morning mist often envelops Thomson's gazelles
as they graze in thousands on the rolling plains near the
Gol Mountains in eastern Serengeti.

# Contents

# Location Map

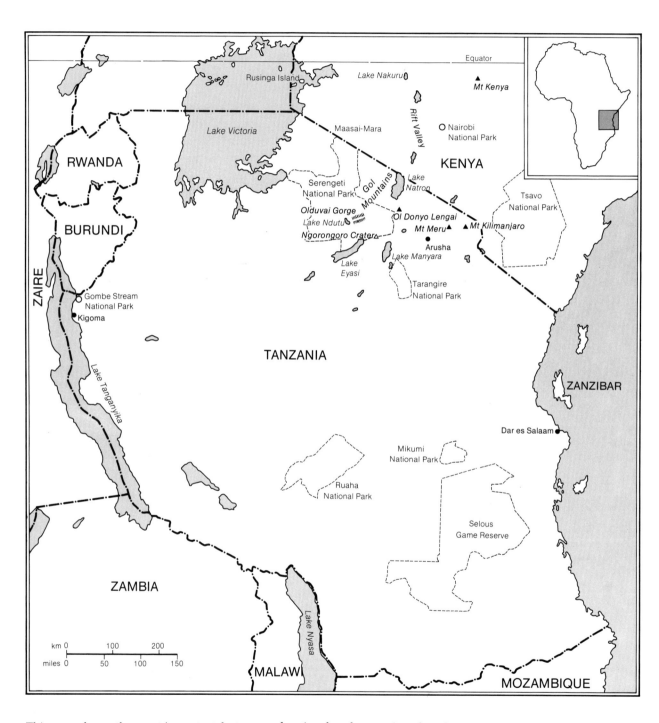

This map shows the most important features and national parks mentioned in the text.

6

# *Photographer Extraordinary*

The first thing Hugo did after driving the Land-Rover up to his favourite camping ground in the Gol Mountains was to walk over to a flat-topped acacia tree and examine a fork in its branches. When, a month or two before, he had camped on the same spot, some empty bags had been lodged in the fork to keep them out of harm's way. On breaking camp and removing the bags he had discovered beneath them the beginnings of a dormouse's nest. So Hugo had left scraps of banana and pawpaw in the fork and fixed a little roof of bark to shelter the incipient nest. Now he wanted to see how the dormouse had made out. No nest, no food, no dormouse. A predator? Or had a family been born and weaned? Impossible to say.

Driving along a rocky, pot-holed track filmed over with powdery dust, Hugo suddenly braked, reversed, stopped and jumped out. I had seen nothing, but my eyes are unpractised and his almost miraculously quick. A small chameleon, dragon-like in his livery of black and gold, was

Hugo spent many years studying and photographing the rarest of the large carnivores in Africa, the wild dogs, of which probably less than 150 exist on the Serengeti, an area larger than the Netherlands.

advancing with measured dignity into the stringy grass which pressed in upon the track. Hugo wanted to make sure that the Land-Rover's wheels hadn't injured the little lizard.

Such trivial incidents illustrate one facet of a contradictory character in which an almost obsessive abhorrence of giving pain, or even causing inconvenience, to any living creature, is combined with an admiration for those animals who hunt and kill to live. He has a special liking for those species most generally rejected and despised by men: wild dogs, jackals and hyenas. It is largely thanks to him and to his first wife Jane Goodall that the image of the lowly, sneaking scavenger has been replaced by that of social animals with strong family ties who make exemplary parents, are often bold and resourceful and who lead interesting lives. He even finds good words for the mouse-bird which most people, anyway those with gardens, look on as a pest. He once tamed one. 'The only bird I know who plays,' he said. 'We played a game together, I tapped my hand on the ground and he pranced round and round it, pretending to attack.'

Whether Hugo is patient by nature I do not know. If not, he has had to learn. Anyone can photograph wild animals in a national park where they have become so used to vehicles that they don't run away. Very few people are prepared to follow a small group of animals for perhaps two years or more, and seize those few moments when creature, light and setting combine to make a picture memorable. Hugo seems to follow a few simple rules. The animal must be doing something interesting – playing, courting, mating, fighting, hunting, killing – not just standing there – unless the quality of the light is extra-special. The filtering effects of cloud and dust, cloud shadows and hill shadows, bands of light and shade, threatening storms, sharply lit horizons – these are his materials. There are many frustrations. If the animal does something interesting, even dramatic, and the light is wrong, what might have been the best picture of the year comes to nothing.

Most of the photography is done from his Land-Rover window, which is fitted with a tripod head for the cameras. Hugo is adept at steering with one hand, changing lenses with the other and, at the same time, re-loading the camera and lighting a cigarette. (He is almost, though not quite, a chain-smoker; drinks, I would guess, fifteen to twenty cups of black coffee a day; and takes little alcohol.) He is prodigal in the use of film.

Once when I was with him he had his eye on a tree leaning out over a pool in the bed of one of those rivers that flow in the rains, dwindle in dry weather to a series of shallow water-holes, and give up altogether in a drought. It was a pleasant little pool, frequented by Egyptian geese and by sacred ibises who were thrusting long bills into the water in search of larvae, small frogs and other nourishment. Masked weaver-birds came and went among the thorn-tree's branches, carrying bits of grass to their loosely constructed nests. Nest-building among weaver-birds is a communal activity conducted by the males who, when the nest is ready, flutter near the entrance to entice a female in. This was the scene Hugo wanted to photograph: the black-and-yellow birds hovering with

A male masked weaver bird building an intricate nest. When completed he tries to attract a female by hanging upside down at the entrance, fluttering his wings.

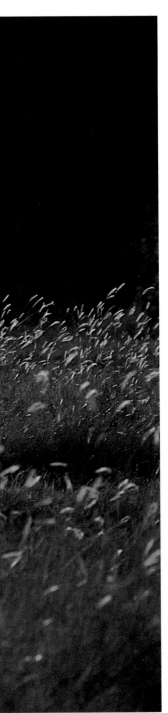

A male impala, the most graceful of antelopes.

sunshine striking through their outspread wings. Tricky photography: the brevity of the exposure limited the depth of focus to only half an inch.

For three evenings running, Hugo sat there from four o'clock till seven with his lens focused on the entrance to a nest. Now and again, not often, a bird hovered in exactly the right spot and the camera clicked. Most of the time it didn't click. The evening was peaceful, the river-bed astir with life. Some ducks alighted, swam around and then took off again, leaving behind a gentle ripple. The harsh cackle of guinea-fowl came from the bush. Two giraffes with a half-grown child were browsing on thorn-tree tops. Suddenly came a sound of hooves thudding on dry, sandy soil overlying rocks: half a dozen male impalas, lyre-shaped horns laid back against their shoulders, leapt with effortless grace, like rufous waves surging over the bushes, as they chased each other in play. Unexpectedly, several elephants strolled down to the far end of the pool, sucked up water, sprayed themselves from their trunks and then retreated with their usual dignity into the bush. The ibises continued their fishing; a flock of sand-grouse alighted by the water's edge; a black-shouldered kite hovered overhead; doves ceaselessly cooed. Above the horizon, the sky reddened as the sun sank towards a steel-blue cloud stained with crimson.

The slanting light threw trees and bushes into dramatic relief, as though a spotlight had been switched on. Long, black shadows scarred the bush. Then the cloud swallowed up the sun, light faded quickly and a grey heron landing by the pool could scarcely be distinguished in the grey dusk. But the sky was still glowing when we jolted back to camp to see four giraffes move with their loping gait along the skyline, their heads and necks black against a smouldering sky. Their motion was like that of boats with masts awry sailing across a gently billowing ocean. They vanished down the slope below the camp and later, in the remnants of vanishing light, could be seen wading across the lake. Soon, frogs would strike up their nightly chorus.

A hoopoe was sitting on a thorn-tree branch beside the track early next morning, handsome in his red-brown plumage and black-and-white banded wings, his crest of black-tipped feathers standing erect. Beams from the rising sun crept gently up the tree-trunk and encircled the bird with a kind of halo – an effect Hugo had aimed at with the weavers. Just right, surely? He shook his head. 'A feather's missing from his crest.'

Perfectionism. We watched the hoopoe till he swooped away with his undulating cuckoo-like flight.

11

Three qualities, I think, are needed in a serious wildlife photographer: patience, perfectionism, persistence. To these must be added contentment with solitude and one's own company. This does not, in Hugo's case, imply aversion to the company of others; on the contrary, he enjoys conversation, relishes the companionship of women and lives on excellent terms with his African staff. While in professional matters a strict disciplinarian, in daily life in camp he is relaxed and easy-going and never seems to be in a hurry. This is the tempo of Africa, and it has crept into his bones.

He has lived now for about eighteen years within the Serengeti ecosystem, which includes the Ngorongoro Crater and its surrounding Conservation Area and part of the Maasai-Mara region of Kenya as well as the National Park. The area covers about 14,500 square miles, larger than the whole of Holland. Will he ever tire of it, I asked – feel that he has squeezed it dry of photographic interest and that it is time to move on? About one hundred species of mammal, he replied, about five hundred

Serengeti means 'an extended place' – an apt description. A semi-desert in the dry season, about four million animals congregate there when the rains have turned the grass green.

species of bird, plus innumerable insects, populate the area. A lifetime is not enough to chronicle them all. Not even for a single species does he consider his coverage to be complete. He does not think it ever will be. The perfect picture lies always just out of reach of the lens.

Baron Hugo van Lawick was born in 1937 on the island of Java, then part of the Netherlands East Indies, now of Indonesia. The barony can be traced back to the thirteenth century. Its coat of arms represents seven bends in the River Linge, supposedly where his forbears, who were Vikings, first landed in Holland. The coat of arms of his mother's family (van Ittersum) features the heads of three donkeys, or possibly mules, for the first mules seen in Holland were brought there by an ancestor at the time of the Crusades. On the other hand there is a story that when the King of Holland, unseated in a tournament by the Baron van Ittersum of the day, exclaimed 'What a donkey I am!' the Baron tactfully replied: 'Sire, in that case I am three donkeys,' and the sally was perpetuated in his coat of arms. There is a military tradition in the family. Hugo's paternal grandfather commanded the Dutch cavalry, and his own father was an officer in the Dutch Fleet Air Arm.

It was while the family was living in Java, when Hugo was no more than two or three years old, that his interest in wildlife may have been born. His mother had a very small pet squirrel who used to make a nest in her hair, covering itself completely and sometimes startling visitors by poking out its head and confronting them with a pair of bright eyes surrounded by fur. Later, it found a mate in the garden and returned to the wild, but Hugo remembers it appearing every day at tea-time to take a lump of sugar from his mother's hand.

In 1941, when he was four and his brother two years younger, his father was killed in an air crash. Soon afterwards, the Japanese invaded Java and the Dutch fled. Hugo's mother and her two small sons reached Australia and spent a year in Perth before proceeding to England by sea. The night after their ship left Sydney, Japanese suicide submarines entered the harbour and blew up the part of the harbour where their vessel had been berthed the previous day.

In England, Hugo's mother married a British officer and the family settled near Tiverton in Devon. Here Hugo's interest in wildlife was fully awakened. In the woods nearby he discovered rabbits, badgers, foxes, many birds and other fauna; butterflies especially attracted him. He must have been lucky in the boarding school (Kestrels in East Anstey) to which he was sent. The teachers encouraged wildlife studies and he gave his first lecture at the age of eight, on the subject of owls.

'I liked school in England,' Hugo recollects, 'but hated it in Holland, maybe because there was very little sport in Holland, the whole atmosphere was one of cramming facts. In England, when I was about nine years old, I had my first boxing match. I had never heard of boxing, and all that was explained to me was that you hit each other and jumped about quite a bit. I misinterpreted the "jumping about quite a bit", so in the ring I just jumped straight up into the air rather like a Maasai dancer, and while I was landing hit my opponent on the nose. Much to my surprise I lost the match, although my opponent's nose was bleeding and mine wasn't. Of course I lost on points. One of the masters offered to teach me boxing in my spare time, which I gratefully accepted, and subsequently I became the best boxer in the school.

'Throughout my school years, I excelled in sports, but not in those that involved teamwork. In other words I wasn't good at football, hockey, baseball and so on, but was good at cross-country running, swimming, fencing, gymnastics and springboard diving. I was energetic and had a lot of stamina, and at the same time was rather shy. If somebody said I couldn't do a thing, I would become incredibly determined to prove that I could.' In this respect at least, he has not changed.

After the war ended, his mother's second marriage having foundered, the family moved back to Holland. Hugo was ten, and stayed at school until he was sixteen. 'We lived in Amersfoort,' he recalled, 'in the house in which my mother was born. It only took about ten minutes' bicycling to get into the country so I spent much of my time out in the woods and meadows. I had a number of pets, including a dog, a cat, two jackdaws, a seal, two foxes, a badger, some grass snakes and some vipers. The vipers I got rid of after I walked into my bedroom when a girl was cleaning the room and noticed that one of the vipers had escaped. I pushed the girl aside just as she was about to pick up my pillow, and there underneath it was the viper.'

'In 1953, many of the dykes in Holland broke in a great storm and I was one of the volunteers who went to help repair them. It was during this time I saved an orphan seal. Like infant apes, baby seals need constant physical contact, and at night I had to sleep on my tummy with the seal underneath my bed, holding on to one of his flippers, otherwise he cried most pathetically. I had the seal for only three weeks; it was impossible to raise him in my circumstances so he was passed on to the Amsterdam zoo.

'When I was fifteen or sixteen, I sometimes joined friends of my mother's when they went out hunting for rabbits, hares and pheasants.

Vultures usually settle early for the night.

But by the time I was sixteen or seventeen, before I went to Africa in fact, I had already turned against hunting, and against having animals as pets. I suppose what might have encouraged this feeling was that I didn't need to prove I was a good shot, because when I was eighteen I was a sharpshooter in the army.'

Academically, his record was mediocre, partly no doubt because of an interrupted education, partly because his heart lay in the countryside and partly because he suffered, he thinks, from a mild form of dyslexia. 'My mother and some of her friends', he wrote, 'frequently expressed worry as to whether I would be able to make a living when I grew up. I think this increased a determination, which to some extent was already there, to prove that I would be able to make a success of life.'

Hugo made his first film when he was fifteen, using an 8mm camera belonging to some school-friends who pre-set the lens and instructed him to crawl to a pre-determined spot, point the camera at a wild moufflon – this was in a national park – and press the button. Photography, he realised, could provide a means of working with wild animals, and working with wild animals was what he wanted to do. Two years' national service had first to be completed. There was some pressure to follow family tradition and stay on in the army. He considered the possibility but turned it down.

Next he had to learn the elements of photography. A company in Amsterdam which made films using puppets took him on as an assistant cameraman. He cleaned and set up cameras, talked to studio crews and picked up all he could about the technique of film-making. Then came his first real break. Armand and Michaela Denis visited Amsterdam. This now all-but-forgotten pair were pioneers in the presentation of wild animals on television. Many of the animals were not really wild, since they lived in their small zoo near Nairobi. But they were exotic, the Denises were good showmen, their cameraman, Des Bartlett, outstanding and their weekly wildlife series *On Safari* was a great success.

Through the enterprise of his mother, Hugo met the famous pair and favourably impressed Michaela with film material he had made on his own account. But there was not much of it. Armand's advice was to get a job in Africa, any job, and learn wildlife photography in his spare time. A year later, Hugo took a chance and followed this advice. After paying for his air ticket, he arrived in Kenya, aged twenty-two, at the end of 1959 with a second-hand 16mm movie camera, some film and very little else. His luck was in. A junior assistant had just left the Denis's employment and he was taken on as a replacement.

Formerly believed to be vegetarians and insect eaters it is now known that baboons often hunt and eat small mammals such as gazelle fawns.

The reality of Africa was even better than the dream. 'Today we took a most wonderful and amusing scene,' he wrote in his diary. 'Two giant forest rats were placed near each other and they immediately stood on their hindlegs, grabbed one another with their forepaws, and started pushing and pulling and squealing loudly.' Two days later: 'When I was filming the spotted hyena, he suddenly charged me from the back, throwing me with my camera straight into the air. He is a very nice animal really, but you have to look out that he does not nip you in the legs.' And on baboons: 'I had always thought baboons uninteresting and very unattractive. Having now had a young baboon for some time and so been able to observe him closely, I have completely changed my opinion. A young baboon is in constant need of company and if left alone will cry as a small child.'

'As he was cold, I took the baboon into bed with me,' Hugo reported in a letter to his mother. 'A baby genet cat has arrived. I carried him around the

17

whole day in my shirt. It took me till after supper until I could really stroke him.' A few days later, the genet was sitting on his shoulder and playing with his ears. 'Behind me the young baboon is screaming because he wants to sit on my lap. He gets hysterical very quickly, and then bites and scratches himself, and gets quite red in the face.'

A few weeks later, the baboon died. Ignorance of animal behaviour was the cause. It is known now, but was not known then, that baboons should not be reared in isolation from others of their kind. Those who receive this treatment as a rule become mentally unstable, attack themselves and sometimes, as in this case, simply die. Commenting on some macaques, Hugo observed: 'All Asian monkeys seem to be homosexual. One even turned around and stuck out his behind when I entered the cage.' Here was another example of ignorance of animal behaviour. Presentation of the bottom is now known to have nothing to do with homosexuality, but to be a gesture of submission, and the mounting of one male by another is a gesture of dominance.

Two young wild dogs also fell victim to this lack of knowledge. One night, while he was dressing, Hugo heard the two animals screaming. Running immediately to their cage he found that it also contained an enormous hyena that he had never seen before. It was too late to save the wild dogs: both had been horribly savaged and had to be shot. The hyena had been brought up on a farm and, as it had never shown any aggression towards other animals, had been wrongly assumed to be harmless, and placed in the same cage as the dogs. Events like this reinforced his aversion towards keeping animals in captivity. 'Continually there were sick animals, accidents, animals dying, and animals brought up as pets put into cages with no time to play with them. Every time you passed the cage one or more of these animals would rush towards you hoping that you would join them and play with them. It was a very frustrating sight.'

Hugo's camera work took him far and wide into the bush, and had its excitements. At this time, rhinos were being captured in areas where poachers were killing them, and were being 'translocated' to the relative safety of national parks. In October 1960 he was camped on a bank of the Tana river where 'the air was filled with the chirping of crickets and smell of wild figs'. A rhino was spotted, Hugo with his camera climbed into the trappers' truck and the chase was on. (This was before the technique of anaesthetising wild animals by shooting a syringe from a dart gun.) 'Very softly the truck crept towards the rhino. Suddenly he looked up and ran

A group of baboons huddle together forlornly during a downpour.

off. Giving full acceleration, we raced after him across country. One of the wheels shot into an aardvark hole and I shot into the air, lost my camera but luckily caught it again before I fell with a big bang on to my tummy in the back of the truck. The rhino took a corner, the truck took the corner too at about twenty-five miles per hour – straight at a thorn-bush. I quickly covered my face, we went over the thorn bush, thorns all around us, and were covered by biting ants.

'The rhino suddenly had enough of it, turned around and charged us. He hit us in the side. We went into the air a little way, getting a hole in the bumper, and the trappers tried to get a lasso around his neck. It didn't work, and the rhino ran off again. Again we raced over holes, over thorn bushes, through dry river-beds and over rocks. Finally we managed to get alongside the rhino and get the lasso around his neck.

'He charges us, a second lasso goes around his neck just in time, because the first one breaks. The rope is pulled shorter, and he can't escape. More lassos are put around his neck, his legs are tied, and then he's rolled over on to a flat piece of wood which is winched up on to the truck. Later this rhino was released in the Nairobi National Park.' He was a lucky rhino to escape injury; many animals succumbed to the crude

Especially vulnerable to predators at night, baboon troops often sleep on steep rocks. Those without suitable rocks within their territory sleep in trees.

catching methods of the day until, in the 1960s, the technique of darting replaced the lasso, and animal catching became a relatively painless business for the animal.

Tame animals could sometimes be as dangerous as wild ones. An acquaintance was killed by her pet lion. 'As so often with people who have tame animals,' Hugo wrote, 'they lose their caution, and she walked past this lion without really looking at it. The lion hit out at her playfully and cut her jugular vein.'

This incident and others like it taught him a lesson essential to survival as a wildlife photographer: never, through over-confidence, to drop his guard. Caution was not native to Hugo's character, but imposed upon it by experience. Occasionally it deserted him. Not lions but caterpillars caused his downfall, in a literal sense. In an attempt to reach a cluster of pupating caterpillars, he crawled along the branch of a tree. The branch gave way, and he ended up in hospital. His case evidently amused the nurses who giggled when they looked at his case-sheet. What was so funny? he enquired. 'Do you really have to run so fast to catch caterpillars?' one of them replied. The cause of the accident had been given on the sheet as 'chasing caterpillars'.

When Armand and Michaela Denis departed on a lecture tour, they instructed Hugo to make camera studies of their zoo animals during their absence. Sometimes these captive animals were placed in a natural setting in order to simulate conditions in the wild. Hugo had developed a strong distaste for this kind of work, so he devoted himself instead to photographing the insects in their garden. It was fortunate that when the Denises returned they praised his insect film instead of sacking him.

After two years, Hugo left the Denis's to set up on his own. Dr Louis Leakey, whose star as a prehistorian was rising fast, invited him to stay at Langata, a suburb of Nairobi. He felt at home there, he has written, with a python in the garden, a tree hyrax in the roof (they screech most of the night) and other animals coming and going or swimming about.

Once again, he was in luck. He was in Louis Leakey's office when a call came through from Washington D.C. The president of the National Geographic Society was in urgent need of a photographer to film the work of Leakey and his wife Mary for use in a lecture tour. He was engaged on the spot. Thus began an association with the Leakeys which led to the making of several films about their finds, and to five years' work for the *National Geographic* magazine establishing him as a wildlife photographer.

This assignment was doubly welcome because palaeontology was one of Hugo's hobbies: he had collected fossils in Norway and in Austria. The day after his arrival at the already famous Gorge he went for a stroll during the siesta hour, spotted a fossil and dug out the skull of an antelope that proved to be about two million years old, and to belong to a hitherto undiscovered species. After this auspicious start he filmed the Leakeys at work at Olduvai, and also some of their discoveries elsewhere. These included a site at Olorgesailie, a hot and arid part of the Rift Valley about thirty miles south-west of Nairobi, where in 1943 Louis and Mary had come upon an astonishing collection of Acheulean stone tools, later pronounced to be about 400,000 years old. With the tools were thousands of fossil bones of extinct animals, much larger than their present counterparts. Pigs the size of rhinos and baboons as big as gorillas had evidently been hunted and devoured on the shores of a now-vanished lake by predecessors of man, belonging to the species *Homo erectus*.

Hugo also filmed sites on Rusinga Island in Lake Victoria where the Leakeys had unearthed the oldest skull of a fossil ape hitherto known, *Proconsul africanus*. Another assignment was a site at Fort Ternan, forty miles east of the lake, where, on a farm belonging to Mr Fred Wicker, nearly ten thousand fossil bones had been discovered, including some that had belonged to miniature antelopes, giraffes and rhinos. In contrast to the Olorgesailie fossils, the rhino here was no larger than a donkey.

It was while Hugo was filming the Leakeys at Olduvai that an exciting find was made by their son Richard, then aged only nineteen but already a keen fossil-hunter. (Mary and Louis had three sons: Jonathan, who after a love affair with snakes settled down on the shores of Lake Baringo to grow fruit and vegetables and organise safaris; Richard, later to become director of Kenya's National Museum and famous in his own right as a palaeontologist; and Philip, the only European to be elected to Kenya's parliament after Independence.) Richard was a few years Hugo's junior and the two young men had become friends. In 1963 Richard Leakey flew over Lake Natron, just south of the Kenyan/Tanzanian border, and observed on the north-western shore, around the mouth of the Peninj River, sediments which looked like promising fossil territory. The question was, how to get there. Few parts of eastern Africa are harsher, more rugged and more difficult to reach. The volcanoes that for millennia have spewed out their lava have created a wasteland of jagged rubble, sun-roasted boulders and sterile ash, mostly too hostile for even the toughest Land-Rover to overcome. 'Touch a tree and it is thorny,' wrote Leslie Brown, the Kenyan ornithologist who pursued flamingos there, 'sit in the shade and you will sit on a thorn, if not on something worse.'

However, they were lucky: on the northern side of the lake there was a rough track which led from Lake Magadi in Kenya. Richard and Hugo's

A mother giraffe and her young calf cross the Serengeti, east of the Gol Mountains. In the background can be seen the live volcano, Ol Donyo Lengai.

Overleaf: A small soda lake on the Serengeti surrounded by fever trees and desert dates.

plan was to camp nearby, build a raft, equip it with an outboard motor, and chug across to fossil-hunt on the other side. A simple plan but, as they soon realised, a rash one. Most of the lakes lying in a chain along the Rift Valley's floor are strongly alkaline, and Lake Natron has an especially high concentration of soda salts, crystallised into an evil-smelling, claret-coloured crust, blotched here and there by patches of green slime and almost impossible for a raft to negotiate. Almost but not quite, for the raft did manage to struggle across while the hearts of its passengers were in their mouths, fearful lest at any moment the engine would splutter to a stop and fail to start again. Their fear was not merely of being marooned in the lake but of being killed, and killed painfully, by the soda. Leslie Brown, in an attempt to solve what he called the mystery of the flamingos – where they breed – had earlier tried to walk across a tongue of the lake, believing that the pink soda crust would support him. It didn't. With each step his feet sank deeper into black glutinous mud and with each step he

grew more and more exhausted in the furnace-fierce heat until, struggling back, death by drowning in the caustic slime seemed inevitable. Utterly spent and dehydrated he just, and only just, made it to the shore. Only luck and the reliability of their outboard motor saved Richard and Hugo from an equally unpleasant experience.

The sand-flats near the Peninj River's estuary were, Richard decided, worth detailed investigation, and they set to work to clear an airstrip with the help of recruits from the local Sonja tribe. After this was completed and equipment flown in, an archaeologist from Olduvai, Glynn Isaacs, began a thorough survey, and almost immediately one of his trained African assistants made a great discovery – the jaw of a hominid over two million years old named *Zinjanthropus*. Mary Leakey had unearthed the skull of a similar creature a few years before at Olduvai.

A few years earlier, Louis had taken on as an assistant at the museum in Nairobi a young girl whose reasons for going to Africa were the same as Hugo's – she wanted to live and work among wild animals. After she had spent some time at the museum, and then at Olduvai, Louis had decided that she was the person he was looking for to make a study of a population of chimpanzees living in a small reserve on the eastern side of Lake Tanganyika. He believed that their environment, the fringes of mountainous forest, was probably very similar to that of the early hominids who had emerged from thick tropical forests to forage in more open, if still tree-clad, country bordering the plains. Observations of their behaviour might therefore throw light on the behaviour of the early ancestors of man, since chimpanzees are more nearly related to *Homo sapiens* than any other species of ape. (These belonged to the eastern or long-haired variety, *Pan troglodytes schweinfurthii*.) The name of the girl was Jane Goodall, and the chimpanzees lived in the Gombe Stream Reserve, about eight hundred miles west of Dar es Salaam, Tanzania's capital. In the summer of 1960 Jane Goodall and her mother Vanne, together with their cook Dominic, established their camp in the reserve near the lake shore, and Jane embarked on what was to become the most thorough and sustained study ever undertaken of a group of wild animals.

Chimpanzees are large, strong, and potentially dangerous animals. At first they were naturally suspicious of this hairless, upright creature, armed only with a tape-recorder, wandering about their mountainous terrain and often sleeping on the ground under a plastic sheet instead of in a leafy nest in the trees. Every attempt she made to approach them resulted in their disappearance into the undergrowth. But gradually their suspicions began to be allayed. There came a moment, at the end of a long,

Three young chimpanzees at play in the Gombe National Park, Tanzania. When fully grown they can be three to four times as strong as an adult man.

frustrating day, when Jane spotted four chimpanzees feeding in the branches of a tree. She carried out a careful stalk to reach a large fig tree from whose shelter she hoped to observe them unseen. Cautiously, she peered round the fig tree's trunk. Yet again, the chimpanzees had vanished. 'The same old feeling of depression clawed at me,' she wrote. But then, less than twenty yards away, she saw two male chimpanzees sitting on the ground, staring at her intently. She gazed back, waiting for the expected flight. But this time the chimpanzees stayed put – and then began to groom each other, a sign that they felt at ease. They were so close she could almost hear them breathing. 'Without the slightest doubt,' she wrote, 'this was the proudest moment I had known.' She knew them both by sight, and had already named them David Greybeard and Goliath.

David, with a distinctive fringe of grey hair around his chin, was the first of the chimpanzees to accept her without fear as a fellow creature. It was to him that she owed the first two of her discoveries about chimpanzee behaviour. One day she saw him with the body of a baby bush-pig in his hands, tearing at the flesh and sharing it with a female and a youngster. This was the first clear evidence that chimpanzees hunt and kill small mammals, and do not stick strictly to a vegetarian diet, supplemented sometimes by insects.

A fortnight later, she saw David Greybeard squatting by the red earth mound of a termite nest. 'As I watched,' she wrote, 'I saw him carefully push a long grass stem down into a hole in the mound. After a moment he withdrew it and picked something from the end with his mouth. I was too far away to make out what he was eating, but it was obvious that he was actually using a grass stem as a tool.' In fact, he was eating termites that had bitten onto the end of the stem.

What exactly is a tool? The *Shorter Oxford English Dictionary* defines it as 'any instrument of manual operation', so a grass stem clearly qualifies. This was not the first or only observation of the use of simple tools by animals. The Galapagos (woodpecker) finch digs insects out of branches with a thorn; chimpanzees in Liberia have used rocks to fracture palm nuts. But these Gombe Stream chimpanzees carried their tool-using a stage further: Jane observed them stripping twigs of their leaves, the better to insert them into termite holes. 'This,' Jane wrote, 'was the first recorded example of a wild animal not merely *using* an object as a tool, but actually modifying an object and thus showing the crude beginnings of tool-*making*.'

Telegrams went off to Louis Leakey, who was wildly excited. Tool-making had hitherto been generally accepted as an attribute unique to mankind. While pulling leaves off twigs to fish for termites may seem a

By altering the shape of a natural object, chimpanzees show the crude beginnings of toolmaking. Here Flo has inserted a grass stem into a termite mound to extract soldier and worker termites. © *National Geographic* magazine.

far cry from the skilled shaping of stone arrow-heads and scrapers, the line which divides the two is a very fine one, if indeed it exists at all. Louis Leakey, with his flair for publicity, made the most of all this and aroused worldwide interest in Jane's observations. Hitherto, the only financial backing Louis had been able to secure for the work at Gombe Stream had come from a Mr Wilkie, the owner of a factory in the United States that manufactured tools of various kinds. He had set up a small museum of tools, including prehistoric ones. The suggestion that chimpanzees used tools had aroused his interest and prompted him to donate $3000 towards financing the research. This provided Jane with a tent, a small boat, her air ticket, and enough to keep her going for six months. Now the National Geographic Society also displayed a lively interest and serious financial worries were at an end.

One of the many theories advanced by Louis was that the early forms of man started their long climb up the evolutionary ladder not as hunter-gatherers, as is generally assumed, but as scavengers. Louis argued that man was, and is, a very slow mover compared with four-legged animals, and that before he invented weapons that killed at a distance he could never have caught his meat; nor could he have torn the animal open with his bare hands before he invented stone tools with which to rip the hide. Hugo recalls that when, as a youth devoid of any scientific training, he had lodged with the Leakeys, he had ventured to challenge the maestro's theory. He advanced the counter-theory that early man and proto-man's skill and cunning would have enabled him, despite his slowness, to ambush his prey; and that the strength of his hands and teeth would have been sufficient to tear his victim apart. Looking back, Hugo is amazed at his own temerity. Louis did not appreciate a challenge to his views, especially no doubt from an untrained, unqualified contender more than thirty years his junior. When contradicted, he was liable to shout. Hugo shouted back. The shouts had reached a crescendo when Hugo suddenly realised the absurdity of the situation. Here was a young, unfledged photographer arguing with a scientist of world-wide repute. He stopped shouting and burst into laughter – good-naturedly joined by Louis. Jane's observations of a chimpanzee tearing a small mammal apart with its hands reached the Leakeys not long after this argument. Hugo made no comment, and Louis had forgotten it in the excitement generated by the Gombe Stream discoveries, and by the vindication of his faith in Jane's potential as a scientist.

Now that her work at Gombe Stream had aroused such interest, especially in the United States, a photographic record was clearly needed. Jane had hitherto resisted proposals to send in a photographer. The

Although chimpanzees sometimes hunt and eat baboons, the young of both species often play together. Such observations lead to interesting speculation of the relationship and behaviour of early man with his prey species before the advent of weapons.

chimpanzees had accepted her, but a newcomer clambering about with a lot of unwieldy equipment might well be a different story. Louis had arranged for Jane's sister Judy, equipped with cameras, to go to Gombe Stream for three months, but continuous rain had literally washed out her endeavours. This continuous rain was no joke. Torrential storms would blot out the landscape for hours on end while lightning flashed and thunder rolled overhead. Humidity was intense and very enervating, and fertility got out of hand; grass grew up to twelve feet high and Jane, soaked to the skin day after day, had continually to climb trees to observe her quarry. As for the chimpanzees, they had not acquired the skill of building shelters and sat hunched in trees or in the open with bowed heads until the deluge was over. Best off were the infants, who sheltered beneath their mothers' bodies and kept dry.

The National Geographic Society wanted a film of Jane and her chimpanzees; Hugo was available, having finished his film about the Leakeys' work, and he had earned Louis' respect as a photographer. Early in 1962 he arrived at Gombe Stream. At that stage only one chimpanzee, David Greybeard, was bold enough to visit the camp and accept bananas. The others could be filmed only from a distance, with a powerful telephoto lens. Their eyesight was so keen that if Hugo approached to within about five hundred yards, they were off. But on his second day he had a stroke of luck. With a powerful 600mm lens, he was able to film several chimpanzees eating a red colobus monkey. Then luck deserted him. In her book *In the Shadow of Man* Jane wrote:

> 'Poor Hugo – he lugged most of his camera equipment up and down the steep slopes himself. He spent long hours perched on steep rocky hillsides or down on the softer earth of the valley floors that always seemed to harbour biting ants. Frequently no chimps came at all; when they did, they often left before he could get a foot of film.'

Gradually the chimpanzees gained confidence and came to accept Hugo as they had accepted Jane. In time, they displayed too little shyness rather than too much. One day, Jane wrote:

> ' . . . when Hugo was crouched up in a small hide by a large fruiting tree, a group of chimpanzees climbed up and began to feed. Just as he was beginning to film, he felt his camera being pulled away from him – then he suddenly saw a black hairy hand pulling at an old shirt which he had wrapped around his camera to try to camouflage the shiny surface of the lenses. Of course it was David Greybeard . . . Hugo grabbed hold of one end and engaged in a frenzied tug-of-war until the shirt finally split and David plodded off to join the group in the tree with his spoils in one hand. The other chimpanzees watched these proceedings with apparent interest, and after that they tolerated Hugo's filming even though, during the struggle with David, most of the hide had collapsed.'

After this episode, the group that included David Greybeard often invaded the camp in search of bananas, and made free with anything they found there, ultimately even allowing Jane to groom them as if she had been another chimpanzee. Subsequently, she realised that this intimacy was a mistake. The strength of an adult male chimpanzee is said to be about four times as great as that of a man. The chimpanzees began to take liberties, and a rule was made, and strictly adhered to, banning physical contact with the animals. If a chimpanzee approached too closely, the human had to move away and keep out of reach. In the early days of Jane's research, she had several times been threatened by angry or suspicious males, and had once been hit on the head with a stick. No chimpanzee had actually pressed home his attack, but that was not to say that no chimpanzee ever would. Among most apes, the upright posture is one of aggression, so perhaps the humans at Gombe Stream won their immunity by walking on two legs. It is also probable that these wild chimpanzees, never having tried physical confrontation with a human, believed the upright walking invaders of their territory to be stronger than they. It is tantalising that we shall never know what went on in the minds of the chimpanzees when confronted with these hairless strangers.

Gombe Stream is a very small reserve – now a national park – only about ten miles long by one to three miles wide enclosing about twenty square miles, and very mountainous. In it dwell three separate communities of chimpanzees, each of which guards its territory jealously; males will fight if those of one community infringe the territorial rights of another. The group Jane was studying numbered about fifty individuals, and individuals they were, each with a separate personality. David Greybeard was bold, self-confident and the first to lose his fear of humans; his friend, perhaps brother, Goliath, was the dominant male. Among the females, Flo and her family fascinated observers most. An ugly elderly matron with a bulbous nose and tattered ears, she was endowed with prodigious sex appeal: in fact she was a real tramp. Whenever she came into season, signified by a swelling and reddening of the sexual area, the males pursued her with unflagging ardour and mated her again and again – a perfunctory business, each mating lasting for only ten or fifteen seconds. There were no signs of jealousy among her suitors but her daughter Fifi, aged three and a half when the study got under way, tried continually and unsuccessfully to interfere and push the males away. (Flo had three children who still kept close to her: Faben, an adolescent of about eleven; Figan, just attaining puberty, aged seven or eight; and Fifi the daughter, still suckling and sharing her mother's nest at night.)

Males play no part in the upbringing of the young, who stay with their mothers until they are fully adult and even then will often return to her for short periods and come to her defence in the case of squabbles with other chimpanzees. Whereas the normal period of oestrus, which occurs about every thirty-six days, is not much more than a week, Flo remained on heat for five weeks, with one gap of about three days. No wonder she looked utterly exhausted. No wonder, either, that she conceived.

Meanwhile, much had been happening among her human observers. Jane had been admitted as a graduate student to Cambridge University –

one of the very few individuals ever to be accepted as a candidate for a higher degree who had never sat for a lower one. Hugo had completed his first film on chimpanzees, which was to be shown to members of the National Geographic Society in Washington D.C. And Jane and Hugo had fallen in love.

They were married in London in the spring of 1964. The wedding cake supported a model of David Greybeard, and portraits of other chimpanzee personalities looked down upon the guests. Completely in character, the pair cut short their honeymoon to hurry back to Gombe Stream, for news had reached them that Flo had borne a son. By the time they reached the camp he was seven weeks old and very tiny. 'His small, pale, wrinkled face was perfect,' Jane wrote, 'with brilliant dark eyes, round shell-pink ears and a slightly lopsided mouth, all framed by a cap of sleek black hair.' Flo showed him off with pride. He was named Flint.

It is easy to understand the spell that these apes cast over their observers even when seeing them at second-hand, on film; and films, of course, will show you aspects of their behaviour that could not be seen in a twelve-month of observation of the living creatures. I think their gentleness is what struck me most. Of course they can be savage,

A female chimpanzee intervenes aggressively when a potentially dangerous adult male baboon, with canine teeth as large as a leopard's, tries to play with her defenceless infant.

especially the males who will thump tree-trunks, hurl great branches about and utter ear-splitting sounds that Jane has called pant-hoots, sounds that can send a frisson of fear up and down the spine. When territories are in dispute it can be a fight to the finish. But in their domestic lives they touch each other gently, clasp hands, give little hand-taps, kiss each others' lips, groom coats with nimble fingers, rest a protective arm on another's shoulders. They seem to be in constant need of reassurance through physical contact.

Their hands, I think, fascinated me most. The fingers are long and supple, the hands themselves narrow and not in the least coarse or clumsy. In one shot a chimpanzee lifts the hand of a companion to his mouth and gently kisses its back just as if he had been an eighteenth-century aristocrat greeting Madame la Comptesse in a Parisian salon. They communicate much more by gesture than by sound. The temptation to be anthropomorphic is hard to resist when animals behave in so human

Among chimpanzees physical touch is vitally important in their relationships. Worried by the arrival of others an infant among a grooming group stretches his hand to an adult male, who may kiss it reassuringly.

a manner. Such was the moment when David Greybeard took a palm nut from Jane's hand and gently pressed her hand within his own in return. Such was the behaviour of Flint, a spoilt child, unusually dependent on his mother Flo and still sharing her nest when he was eight years old. By then he was physically quite capable of leading an independent existence but seemed unable, or at any rate unwilling, to sever the emotional bond. One day Flo, by then old and weary, disappeared. Jane and Hugo searched and found her body lying half in and half out of a stream she had been trying to jump. Flint was nearby. He never left her body or its near proximity, scarcely ate, and after three weeks he died.

The chimpanzee's eyes are its other most human-seeming feature. They are dark and deep and sad. The sadness, however, is no doubt in the eye of the beholder, for the free chimpanzee is on the whole a cheerful creature. Perhaps human imagination invests them with a foreknowledge they cannot have of sorrows that may lie in store. At present they are well protected in their tiny national park but, as everywhere, the human tide presses in to threaten their habitat. Not so long ago, a chimpanzee might almost have swung across Africa from the west coast to the Great Lakes, the eastern boundary of their range. Now the chimpanzees at Gombe Stream, and another community near the southern end of Lake Tanganyika, are virtually vestigial populations.

The simple camp at Gombe Stream was meanwhile growing into a sizeable village. Research was attracting students of animal behaviour from many parts of the world, and soon fifteen to twenty young men and women were at work there, requiring to be fed and housed and cared for generally. Hugo took over the administration of a complete research centre, no easy task. In addition to the large amount of paper work involved, there was also a sizeable Tanzanian staff to be looked after. Supplies had to come by boat from the small town of Kigoma, often cut off by floods from the rest of the country for months at a time. Communications were difficult and at times non-existent. Hugo's reputation was growing – and so too were his ambitions. Jane's mother Vanne had these observations to make:

'In those days he was intolerant of young assistants who were not prepared to work for him with the same dedicated zeal as he worked for himself. One poor girl, finding that typing in the afternoons in a tent in a temperature of about 100 degrees and a very high humidity was almost intolerable, used to absent herself for an hour or so. One day Hugo saw that she was not at work in her tent. Several days later he came across her retreat. She had dammed up a small pool in the icy mountain stream just beyond the camp and foolishly left her footprints and her towel on the bank. Hugo was furious. The incident sparked off a major camp row.

On the other hand Hugo could be generous and compassionate. I had been typing a manuscript under the shade of a small banda (temporary hut) near the tents when a strange chimp approached the camp. When this happened the rule was – absolute silence – all humans into the tents which must then be speedily laced up. In my hurry I left my MS on the typewriter. The stranger chimp came into the camp. Only the forest sounds could be heard above the rushing of the stream. The heat in the tents was stifling. The chimp moved silently to the banda, picked up my MS and began to chew. A corner was eaten. It was my first attempt at writing a book and I didn't think I could ever tackle those chapters again. I happened to be in Hugo's tent. I gasped and pointed. The chimp went on chewing.

Suddenly Hugo, throwing all rules to the winds, snatched a pair of well worn socks and hurled them towards the vandal. The chimp paid no attention. Hugo's best tie followed and then a shirt and finally a towel. The chimp took fright at last and hurried to a nearby hillock, leaving a trail of half chewed paper behind him. Hugo followed. Ten minutes later he returned with a broad grin. He had rescued every sheet of the book.'

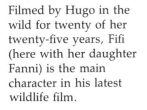

Filmed by Hugo in the wild for twenty of her twenty-five years, Fifi (here with her daughter Fanni) is the main character in his latest wildlife film.

Freud, Fifi's eldest son, relaxes on the ground in the late afternoon.

Students at the camp would be out all day following chimpanzees through thick forest and up and down hillsides so steep that they had to hang on to roots and rocks as they climbed. There was one tragedy. A girl who had been following chimpanzees up and down mountains was looking up into the trees as she walked, and failed to see a precipice at her feet. She tumbled down and was killed. Her body was not found for six days.

Several years later, in 1975, the peace of the Gombe Stream Research Centre was shattered by another traumatic event. Around midnight, a party of about forty armed men landed from motor boats, raided the camp, and kidnapped four students; two American girls, a Dutch girl and an American youth. Jane and her small son were there, but in a hut a little way apart from the rest of the camp, and they escaped notice. The raiders had come across Lake Tanganyika from Zaire on the opposite side. They claimed to be members of a rebel army that was carrying on a guerrilla war against the country's armed forces under the command of its president, General Mobutu. Their aim was to support the separatist cause of the southern province of Katanga.

Thus Jane and her colleagues found themselves immersed in the unsettled waters of post-Independence African politics, and a focus of world attention. Not long after the kidnapping, two men walked into the American embassy in Dar es Salaam, demanding a ransom for the abducted students. After several weeks of acute anxiety the affair ended happily: all four students were released unharmed. But the after-effects were long-lasting. The Tanzanian Government ordered the withdrawal of all foreign students, and that was the end of the Gombe Stream Research Centre in its original form. A much reduced camp has since been manned by Tanzanians only, and a number of trained African workers continue to follow the fortunes of the chimpanzees under Jane's supervision.

So the chimpanzees have not dropped out of sight. David Greybeard, Goliath and others of the group are dead – the lifespan of a wild chimpanzee is about forty years – but Flo's daughter Fifi is alive and flourishing, and has three children of her own – Freud, Frodo and Fanni. In 1982 Hugo returned with his cameras to round off the story of Flo's family, the central theme of a major film covering twenty years in the life of this chimpanzee community. The film will take its place as the most complete and perceptive photographic study ever made of the life of wild chimpanzees – indeed of the behaviour, interacting relationships and fortunes of any community of wild creatures in their natural habitat.

# Lake Ndutu

After a couple of years or so at Gombe Stream, Hugo felt the need to look for subjects further afield. Administration was taking up much of his time, but it was not advancing his career as a photographer and this was worrying him. 'No one could be better company when things were going well with his work,' wrote a friend, 'and no one more cast down when things went wrong. At such times he would lie on his bed smoking, planning his next move, fanning the flames of a grievance. At this time he was jealous of his growing reputation as a top-ranking wildlife photographer, often wary of business offers, stubborn in his refusal to do less than his best . . . and his best gradually became very good indeed.'

In pursuit of new subjects he made several trips to the Serengeti National Park and the Ngorongoro Crater in north-eastern Tanzania. By then the late Billy Collins (who became Sir William) was urging Jane to write a book about her chimpanzees. Before embarking on that, however, it was agreed between the three of them that Jane and Hugo should jointly write about the predators and scavengers of the Serengeti, and that Hugo should photograph them. Subsequently this was narrowed down to three of the lesser studied species; hyenas, to be studied by Jane, and wild dogs and jackals, allocated to Hugo. The result was the best-selling book *Innocent Killers*, and three films about these lesser and much maligned African predators.

By this time Hugo junior, known to everyone as Grub, had made his appearance. He was born in Nairobi in 1967. His parents decided that, from the start, he would go with them wherever they went, which to begin with was the Ngorongoro Crater, and that, as with chimpanzees, there would be constant physical contact between infant and parents, and breast feeding on demand. His cradle was a Volkswagen bus, from which Jane continued her study of hyenas and on whose sides, in order to keep at bay inquisitive tourists, was painted in large letters: RESEARCH – PLEASE DO NOT DISTURB. This served its purpose in the Crater but caused considerable hilarity when the bus was parked, with curtains drawn, in Nairobi. Grub grew up with lions, hyenas, wild dogs, chimpanzees and other creatures if not exactly as companions, then as familiar and, in general, friendly fellow inhabitants of the bush. He could imitate sounds made by wild animals before he could speak. Missing out on infant prattle, he spoke from the very first in grown-up sentences. Hugo recalls the first sentence he heard his son speak. 'Big lion out there eat me,' said Grub and roared with laughter.

It took Jane and Hugo over two years to collect the material for *Innocent Killers*, though not two uninterrupted years: they returned at intervals to Gombe Stream, a bumpy and sometimes flood-obstructed journey of eight

Golden jackals live in pairs and may remain together for life. Although cooperating fully in caring for their offspring and when hunting larger mammals, they seldom seem to be able to feed together very peacefully.

hundred miles, to keep in touch with the chimpanzees and supervise the research. (They were also in daily contact with the research centre by radio.) During their absences they arranged for student assistants to carry on with the observations so that a continuous record was maintained.

Neither of the partners had started out as a trained scientist; Jane became one, gaining her Cambridge PhD, but Hugo entertained no academic ambitions. Nevertheless, a serious wildlife photographer must employ techniques very similar to those of the ethologist, or student of animal behaviour. He must study his chosen animal, or group of animals, for days, weeks and months on end from dawn till dark, and at night as well if they are nocturnal hunters, noting down their every movement,

Young spotted hyenas often have a chance to play with their peers.

their relations with each other, their diet, their breeding pattern, everything about them. In order to be at the right place at the right time he must even learn to read their minds so as to anticipate their actions. He must know, for instance, when and how a predator will make its kill. Then, when the long day's work is over, in the evening he must write up his notes and check his findings in the cramped conditions of a safari bus. It is not a life for the easy-going nine-to-fiver.

Like everyone else in Africa, I was brought up to believe that hyenas were nasty, cringing, crafty scavengers living partly on scraps from kills made by others, such as lions, and partly on refuse and human corpses. Certainly they devoured corpses. It was the custom in those days among certain African tribes to drag a dying person from his or her hut into the bush where he or she was quickly finished off; if someone died inside a hut that hut had to be burnt down. Hyenas had been known to carry off unattended babies and their powerful jaws could crunch almost anything. I remember hearing that, in the First World War, troopers in the East African Mounted Rifles, sleeping out in the bush, would use their saddles as pillows in order to keep them safe from marauding hyenas, who would certainly have attacked the troopers too had it not been for sentries and fires. Hyenas were feared as well as hated, and to call a man a *fisi*, their Swahili name, was a deadly insult.

No one had seriously challenged the belief that hyenas scrounge their meat instead of killing it until Dr Hans Kruuk, a scientist based on the Seronera Research Institute in the Serengeti Park, made a thorough and prolonged study of the spotted hyena, *Crocuta crocuta*. This proved the animal to be a powerful predator in its own right, able to hunt and pull down zebras, wildebeests and other large animals as well as smaller prey. (It is a versatile feeder, and will sometimes even catch fish.) On occasion, when gathered in a pack, hyenas will even drive a lion or a leopard off its kill. Dr Kruuk also revealed the clan structure of the species. Hyenas live singly or in pairs for much of their time but will also gather in packs to hunt and to defend their clan's territory. This study, carried out in the Ngorongoro Crater between 1965 and 1970, formed an indispensable basis for future work on spotted hyenas, but Jane and Hugo resolved to make their own independent observations on which to base *Innocent Killers*, a book which did much to change the popular image of the hyena from that of the corpse-eating scavenger to that of an animal which, though far from endearing in many of its habits, nevertheless is sociable and sometimes brave, and whose females are devoted and affectionate mothers. Jonathan

Kingdon, artist and ethologist, whose magnificent *East African Mammals* has become the bible of all students of African mammalian wildlife, has even suggested that there may be similarities between hyenas and our own early ancestors, particularly in such matters as 'fierce group rivalries, the capacity to switch from individual foraging to group hunting, and a long period of exclusive dependence by the young on its mother'. The hyena is indeed climbing up the ladder of public estimation, or at least the estimation of ethologists.

Jane and Hugo displayed in their book a rare combination of scientific accuracy and sympathetic reporting. They did not allow the risk of being accused of anthropomorphism to deter them from entering into the lives of their study animals and presenting them as personalities. They gave them, for instance, appropriate and amusing names, as Jane had done with chimpanzees, and chronicled their doings in such a way as to kindle in their readers emotions of affection, pity and dismay – an unscientific

Formerly believed to be cowardly scavengers, hyenas are now known to be formidable hunters able to tackle quite large animals, including very young rhinos.

way to go on, in the opinion of the more austere scientists. Some such, intent on avoiding all emotional involvement, give their subjects numbers instead of names. A creature known as 17B or 151x4 is probably less riveting to read about than one called Baggage, Jewel or Fifi. Accuracy and austerity are not, however, indivisible. For Jane and Hugo, accuracy was always paramount. They are, wrote Louis Leakey, 'not ordinary animal behaviour students; they have a patience and a persistence in the field that is almost unmatched, and when this is combined with Hugo's own very special qualities as cameraman, the result makes most other animal books seem rather superficial.'

From Ngorongoro Jane, Hugo and Grub moved on to the shores of Lake Ndutu, also called Lagaja, on the south-eastern perimeter of the Serengeti National Park, and just over its border: no permanent camps or dwellings, other than those of wardens and rangers, are allowed inside the Park. It is

Members of a hyena clan regularly patrol the border of their territory during which skirmishes may erupt with patrols from a neighbouring clan.

a pleasant spot. The tents are pitched in a grove of umbrella thorns, *Acacia tortilis*, whose twisted, criss-crossed branches throw a dappled pattern of shade. On the lake below, flamingos stand in close-packed, rose-pink thousands, sucking up algae and keeping up a murmuration as if engaged in perpetual conversation. From this base, Hugo made his film *The Wild Dogs of Africa*. It won him six awards and he considers it, 'I hope without presumption', to be a classic.

He selected for his studies a pack which he named Genghis, after an old dog who was then its leader. He or an assistant (for he had sometimes to be away) kept the pack under observation for about three years, both by day and, as he described in the following account, sometimes by night:

'There was a full moon and I was sitting in my Land-Rover with a crash helmet on my head, my safety belt firmly fastened, and with a thick padding of foam-rubber stuffed all around me for extra protection. I was about to embark on one of my more dangerous but also most exciting jobs. Without using the car lights, I would try to keep up with a pack of wild dogs as they raced across the African landscape in pursuit of their prey. Only the moon would light my way and warn me of the dangers ahead – potholes, rocks, and even waterholes and precipices – for if I used the car headlights I would temporarily blind the hunters and their prey and I did not wish to influence the behaviour or affect the chances of either.

'Outside on the moonlit Serengeti plain, I could see the dim shapes of twelve African wild dogs . . . Suddenly, the dogs were off into the night and I raced after them. At first, it looked as if I might lose the pack, for they quickly left me behind, but as the car gained speed I caught up with the last two dogs and drove parallel with them. They were used to my following and so took no notice, concentrating instead on a Thomson's gazelle which fled in zig zags across the plain ahead of them.

'I glanced at my speedometer. We were going at 30 miles per hour, but the dogs were gaining on me and so I increased my speed. I knew from experience that if I could not keep up, I would probably lose them in the dark and might not find them again for a month or more. This is because a pack normally covers ten to twenty miles in a night, going in any direction, and roams over an area of about 1500 square miles. Searching for African wild dogs is almost as bad as searching for a needle in a haystack, and so once I did find a pack, I tried hard to keep up with it. That night, as I raced across the moonlit plain, I kept a close

A pack of African wild dogs consists, on average, of about ten members.

eye on the two dogs running next to me. As a result I did not give enough attention to the ground ahead, and suddenly, without warning, the front of my car bucked violently sideways into the air and then crashed to the ground again. In spite of the foam-rubber, I banged my shoulder hard against the side of the car and at the same time the steering wheel spun out of my hands, twisting my thumb as it did so. Then one of the rear wheels hit the pothole and the car came to a shuddering halt. At some stage, I had jabbed my feet onto the clutch and the brake and so the engine was still running. I cursed, quickly put the car into four-wheel drive and first gear, and tried to climb out of the hole, but the back wheel spun and dug itself in deeper. I threw the gear lever into reverse, accelerated and let the clutch go. The car shot out backwards. As it did so, I quickly turned the steering wheel slightly so that both front wheels missed the hole. All this had only taken seconds and I could see the dogs vaguely but far away. I raced after them, and soon I was going at 40 miles per hour and catching up, but twice I was delayed as I had to swerve around two more potholes. Finally, I caught up, but almost immediately I came to a screeching halt amid a cloud of dust. The dogs had suddenly caught the gazelle and were tearing it apart, making strange bird-like twittering sounds as they did so.

'They had not been feeding for long when two larger shapes, attracted by the sounds, came racing across the plain – two spotted hyenas. Boldly, the hungry hyenas dashed in among the dogs and tried to steal some meat, but moments later they were giggling nervously as the dogs surrounded them, and then they growled and roared and screamed as the dogs bit at their bottoms. Moments later, the two hyenas retreated and the dogs continued to feed. Within the next five minutes, however, more and more spotted hyenas appeared, arriving in ones and twos from all directions, and finally the dogs were outnumbered and had to leave the last scraps of their kill to the hyenas, which started to approach with closed ranks, whooping and growling as they did so.'

Days were normally less strenuous, often spent watching the dogs playing, sleeping, scrapping, suckling their pups and in other ways passing the non-hunting hours near the mouth of the den. A hierarchy was strictly enforced and Havoc, a particularly ruthless individual – coal black, with red eyes – was the dominant female. Dominant is indeed the word; she bullied the other females without mercy and when one of them, Black Angel, gave birth to a litter of pups, one by one she deliberately killed them in spite of their mother's desperate efforts to conceal and protect them. At last only one, the runt of the litter, survived. Hugo named her Solo.

The courage and sheer guts of this half-starved, spurned and puny pup were quite amazing. Wild dogs seldom stay in one place for long, and whenever the Genghis pack set out, this tiny pup tottered after it on her weak legs, growing more and more exhausted, lagging far behind, but never giving up. When it became obvious that she could not survive for more than a few more hours, Hugo broke the rule observed by all students of animal behaviour, never to interfere with the course of nature, picked up the exhausted pup and took her back to the camp, where Jane slowly nursed her back to health.

When Solo had recovered, the problem that arises in all such rescue operations confronted Jane and Hugo: what to do next. Hugo searched every day for four weeks for the Genghis pack hoping, without much confidence, that its members would accept her back. He failed to find

It is a fallacy that African wild dogs take much longer to kill their prey than do the large cats, although it looks more gruesome. It is also questionable whether the prey feels much – in humans, for example, deep ripping wounds are often not felt for at least ten minutes. A wild dog's prey is usually dead within three and a half minutes.

them. But he did find one of the subordinate females, Lotus, who with her mate Rinogo had left her companions to bear and rear her pups. Unlike chimpanzees, male wild dogs seem to recognise their paternity and will often accompany their mate when she leaves the pack for her *accouchement*, perhaps with the intent of saving her pups from the fate meeted out to Black Angel's by the ruthless Havoc. When the pups are reared, the male returns to his own pack. Hugo spotted Lotus with a litter of pups playing round the mouth of a den. It seemed most unlikely that she would accept this stranger smelling of humans, but against all the odds she did. Solo was left playing with her adoptive siblings at the mouth of their den.

The average litter of an African wild dog is ten but one female on the Serengeti did give birth to sixteen pups. The markings on each dog differ, presumably as an aid to recognition.

What became of Solo? Her saviours had become strongly attached to her and time and time again, during many years spent on and around the Serengeti, Hugo searched for her and for Lotus. He never saw either of them again.

When Hugo came to make his wild dog film he centred it on the story of Solo. The film brought forth from viewers a large number of letters, now preserved in a file. Some are full of admiration, but more are full of shock and horror, accusing Hugo of callous indifference to suffering, inhumanity and general beastliness on two counts – standing by inactively while Havoc 'murdered' Black Angel's pups; and causing to be shown scenes of revolting cruelty, especially a sequence in which the pack chases, catches up with, and tears to pieces an unfortunate zebra. Such scenes, these angry letters said, should never have been shown, especially at times when children might see them, and Hugo was declared a monster to film them and to find good words to say about such disgusting animals. Such protests do credit to the compassion and love of animals displayed by many members of the public, although perhaps, in blaming Hugo, their writers shared the irrational motives of those ancient Greeks said to have killed messengers who brought bad news. In other words, Hugo was castigated for depicting the truth.

He remains unrepentant. Wild dogs have for long been condemned as ruthless killers who always hunt down their prey and then disembowel the helpless animal and eat it alive. Packs have been destroyed as vermin even in parks and game reserves. Personally I cannot share Hugo's affection for them, no doubt because I have never got to know them at close quarters. I do remember an occasion when a pack of fifteen or twenty surrounded a Land-Rover in which I was travelling and jumped around it as if trying to get in, snapping and uttering whimpering cries not at all like barks. They struck me then as sinister and savage, and I thought that if they got into the Land-Rover, or I got out, they would tear me to bits. This was no doubt imagination; I do not think they have been known to attack an active human. But they were not, to me, attractive animals.

But Hugo springs to their defence. They do not, he says, by any means always catch their quarry – though I don't know why success in doing so should be held against them. Hugo and an experienced assistant, James Malcolm, have between them logged over five hundred hunts and only about half of these ended in a kill. When the dogs do kill, they certainly tear their victim to pieces, as do foxhounds when permitted. This, he believes, may be more merciful, in so far as it is often quicker, than the usual method followed by the big cats, which is by strangulation after they have leapt on their prey. He has timed many kills, and found that as a rule the wild dogs' prey is dead within three or four minutes, although young, inexperienced dogs may take up to fifteen. So may lions and cheetahs. Normally they do not kill more quickly than dogs do, but it looks less messy, since the disembowelling generally (not always) takes place after the animal is dead. Moreover, the senses of the disembowelled animal may be blunted by shock, almost as if it had been anaesthetised. He quotes the famous case of Dr Livingstone who, when mauled by a lion, experienced no pain but a sense of numbness, ordained, in the opinion of the great missionary, by a merciful Almighty to spare the victim suffering.

Dogs hunt to live, like all predators; if they fail they die; they commit no deliberate cruelty. Almost certainly, they mark down for their prey animals that are sick or less swift and efficient than their companions, and thus act as nature's instrument to ensure the survival of the fittest. Their social lives are interesting, affectionate, at times endearing. They are, after all, dogs, and like their domestic cousins each one has its separate personality. Hugo will not hear a word against the wild dogs of the Serengeti.

After the wild dog film, Hugo, still based at Ndutu, turned his attention to the larger predators and made a film about the lions of the Serengeti. The demands of his photographic projects were taking him away from Gombe Stream for longer and longer periods. He and Jane had agreed that she would spend roughly half the year with him in the Serengeti area, and he would spend the other half with her and the chimpanzees. The arrangement didn't work. The Gombe Stream Research Centre with its many students needed Jane's almost constant supervision; nor could Hugo neglect for long the administrative side. And much of the time he spent at Gombe Stream was time wasted so far as photography was concerned. Both were dedicated to their work, and that work increasingly drew them apart. The marriage that had started so auspiciously ran on to the rocks. In 1974, Jane and Hugo were divorced. Both remarried, Jane within the year to Derek Bryceson, director of Tanzania's National Parks and the only European minister in the government.

Hugo spent the best part of the next three years in making a photographic study of the Serengeti's six major predators – lions, leopards, cheetahs, hyenas, wild dogs and jackals – to be published in 1977 as *Savage Paradise*.

Then came another assignment. A young English girl, Stella Brewer, daughter of a forest officer in The Gambia, had started a rehabilitation camp in Senegal where she was restoring to their natural environment chimpanzees who had been illegally captured as infants, or who had been captive-born. Naturally these chimpanzees, reared in totally artificial conditions, had no idea of how to fend for themselves in the wild. Stella Brewer had to teach them everything, even such simple skills as plucking fruit from trees, breaking it open or sucking it, drinking from a stream – even munching termites, and the only way to teach them this was to munch termites herself. Stella was a remarkable girl, as devoted to her chimpanzees in her way as Jane is in hers.

Hugo made his film, *Stella and the Apes of Mount Assirick*, and very nearly lost his life in the process. Among these chimpanzees was a large male, uncertain of temper and suspicious of humans, no doubt with good cause. He threatened Hugo several times, with much more determination than the wild chimpanzees at Gombe had displayed, and finally sprang upon him and sank his teeth in Hugo's neck. He missed the jugular vein by a hair's breadth.

It was while he was in The Gambia and Senegal that Hugo met and married his second wife Terry. Their marriage lasted less than three years. Life in the bush may seem exciting from a distance and be delightful for holidays, but as a permanent arrangement it is apt to prove uncomfortable, sometimes alarming and above all, lonely. In pursuit of his pictures, Hugo would often be away from camp for days and nights on end, sleeping in his Land-Rover, living on bread and butter, hard-boiled eggs and endless cups of coffee, going to bed at nightfall and getting up at dawn. If Terry went too she would sit for hours by day with nothing to do, and by night perhaps hear lions sniffing round the vehicle. If she stayed in camp, she would have no company beyond the staff who only spoke Swahili, no occupation save reading and no defence against night-time fears. Attractive young women content with such a life are rare and Terry was not one of them.

# The Wildebeest Migration

Hugo's camp overlooking Lake Ndutu consists of eight green canvas tents pitched in a row, each with its small veranda; a kitchen and store made of corrugated iron whose roof catches rainwater to be stored underground; nearby, three more tents for his African staff; and two small *choos*, earth closets with a deep hole. There is also a 'shower', consisting of two plastic basins on stands, whose contents you pour over yourself after washing. So, while simple, it is comfortable, having good mattresses instead of camp beds, a library – one of the tents – with easy chairs, and an experienced cook. Hugo rebuts the charge of sybaritism by pointing out that this camp is his only real home.

Sirili, the cook, was formerly in the service of a German movie star, Hardy Krüger, who built a mansion above the forest on Mt Meru. Later, he was taught curry-making in the household of an Indian family in Arusha; his curries are first-class. The local meat, bought in a village thirty miles away in the shape of a cow, is apt to be tough, so curries, with their long slow cooking, are probably the best way to treat it. Sirili, now in his late

Hugo's home and base camp for the past sixteen years on the Serengeti at Lake Ndutu, which means 'a peaceful place'.

fifties, belongs to a dwindling band of Tanzanians who served in Burma in the Second World War. His people are the Chagga, whose country straddles the fertile foothills of Mt Kilimanjaro, where he has a wife and family and a small farm with coffee bushes and banana trees and two cows, quite a prosperous little set-up.

Members of the camp staff have become a close-knit small community. Besides Sirili there is Renatus the mechanic, on whose skill Hugo's survival may depend; Sirili's good-looking young nephew Lawrence; an elderly odd-job man who seems always to be roaring with laughter; these are friends rather than servants. Necessary visits to Europe or America oblige him, on occasion, to leave his camp, possibly for several months at a time. The camp routine ticks over during these absences, so far without trouble or fuss. You cannot lock up a tent. To return, laden with presents, to find everything intact – no pilfering, no defections, such crises as may have arisen through weather, accident or other causes satisfactorily dealt with – is a measure of the staff's reliability.

From the camp under the umbrella thorns you can see below you part of Lake Ndutu, and its flamingos who seem never to pause in their food hunt and in their chatter. There is always something else to see: a group of elands on the far side; zebras grazing right up to the tents; elegant chestnut-and-white impalas with their flickering tails; a pair of dikdiks inhabiting a dell just beyond the trees; and many other creatures going about their daily tasks.

There is nothing rare about the starlings, rightly called superb. They have iridescent sky-blue backs and wings with rosy breasts encircled by a white stripe that puts me in mind of an old-fashioned gentleman's watch-chain. A mother starling fed three children who were just as big as she; they begged with open bills quite shamelessly. These starlings are quick to spot snakes, and will sometimes play the part of an early-warning system. A pair of d'Arnaud's barbets came, stylish in their speckled plumage with yellow rump and patches of red under the tail. Then golden canaries; wagtails; endearing little crimson firefinches; and many, many doves.

Scuttling about among the birds were two small brown rats, more like mice to European eyes, and harmless. Sunbirds with their glittering plumage and slender, curved bills came glancing in, hovering with vibrating wings, and sped away. Their nests, made of spiders' webs, were behind the kitchen. A pair of wire-tailed swallows was endeavouring to build just inside the dining-tent. This was a mistake on their part, as when Hugo and his guests, if any, go out or go away, the tent is zipped up and the swallows can't get in. But they go on trying.

Fischer's lovebirds, a pair of which nests in a hollow tree in the camp.

Four Fischer's lovebirds are in occupation of a hole in a tree which has an interesting history. It was made by woodpeckers, but they were chased off by a pair of Hildebrandt's starlings, who nested there. Then it was taken over by sparrows. The lovebirds are the fourth set of occupants, to date. Holes in the ground, like those in trees, are also handed over, as it were, from tenant to tenant. The pioneer digger is quite often an aardvark, a nocturnal termite-hunter with a long, pointed nose, or perhaps a honey-badger (ratel). Then a hyena may make use of it, or a jackal, or a wart-hog, or a wild dog. Holes may still be in use which were first dug a century and more ago. The plains are riddled with them, to the detriment of shock-absorbers and springs.

Just outside the dining-tent, a hollow log serves as a bird-bath and food-table. Birds come and go continually to sip the water, ruffle their feathers and feast on bananas, pawpaws and other fruits, and on millet grain. One could watch them for hours. Gaudiest of all is the Fischer's lovebird, a small parrot with all the colours of the rainbow in its plumage: an emerald-green back and breast, orange throat and cheeks, a golden neck, a sepia crown, with cobalt-blue tail-tip. I saw six on the bird-table at one time. Formerly common, they are being trapped with bird-lime on a distressing scale, to be sold into captivity for the gratification (presumably) of pet-keepers the world over.

Many kinds of cuckoo inhabit the bush, each specialised to parasitise some particular species. One which lays in the nest of Hildebrandt's starling has set a puzzle. The greater spotted cuckoo in question is too big to enter the starling's hole. Yet the egg gets in. How is it done? The only answer seems to be that the cuckoo lays on the ground, picks up the egg in its bill and drops it into the hole. No one has actually seen this happen. But once, on returning to camp, Hugo found an egg marked like a starling's lying on a table in the veranda of a tent, and opposite the tent was the tree in which the Hildebrandt's were nesting. Had the cuckoo been disturbed after she had laid the egg, and been unable to complete the rest of the operation? Supposition: but possible. Hugo is still hoping to secure a picture of this bizarre behaviour, if indeed the supposition is correct.

Hugo's camp lies a few hundred yards outside the Serengeti National Park, and in the Ngorongoro Conservation Area, where some humans may live but none may cultivate, not even a vegetable plot, nor may they establish permanent villages. But the Maasai may, and do, pasture their goats, sheep and cattle here, and build their *manyattas*. Formerly the Conservation Area, covering some 3200 square miles, was part of the National Park, but in 1959 it was excised and put under separate management to meet Maasai claims.

Many people feared, at the time, that this would spell disaster for the wildlife, but so far it has not. Hunting has continued to be banned, as well as permanent human settlements. In practice, one of the important differences between Park and Conservation Area is the presence in the latter of the Maasai with their abundant livestock. The great Ngorongoro Crater is at its heart. Since the excision, two stretches of land to the north and west have been added to the Park and these have made up most of the lost acreage. But it is a different kind of land, with more bush and less areas of open plain.

Cordon bleus, one of the numerous species which daily visit the bird bath at Hugo's camp.

During my visit the sun rose every morning at twenty minutes to seven. By this time Hugo is normally away in his Land-Rover in whichever direction he has decided to go. This depends on many factors: on news that may have come in of the location of some particular animal or group of animals he wants to inspect; on rain that may have fallen in such-and-such an area; on an animal he has been following who may be due to have her young. My visit coincided with the wildebeest migration, so it was towards the country he believed them to be travelling through that we set forth. First, Hugo checked the lenses and, scarcely less important, Sirili loaded four out-sized Thermos flasks full of hot water into the back, and also put in two cups, a spoon, a tin of instant coffee and a few basic provisions in a plastic container. Also we took binoculars and Hugo's indispensable cigarettes.

We headed south-westwards, through a tract of light bush not so thick as to hide the animals, but providing them with shade and sometimes food. Most of the trees were acacias of various species, interspersed with a straggly, dwarfish little tree with a twisted bole called commiphora. The bloom of most of the acacias was almost over, but one species was smothered as thickly as a hawthorn in England in June with tiny, pale lemon-coloured fluffy balls that had a faint, rather sharp but very pleasant scent.

The *tortilis* species seems incapable of standing upright or imposing any sort of order on its branches. Its trunks lean over at all angles, the branches interlace like so many demented snakes. The flowers come before the leaves and the tree's innumerable twigs and boughs were bare save for minute leaf-buds as yet tightly clenched. A tall *tortilis* stood outside my tent and at night the stars, so much brighter and more thickly clustered than those we see in northern climes, shone through this spiky network in pinpoints of frosty light that brought to my mind, incongruously, a giant Christmas tree festooned with countless candles flickering in the dark.

Many of the trees near the camp were dead or dying. Some stood blackened and forlorn, others lay prostrate in various stages of decay. Some of this damage had been caused by fire but much of it was due to elephants, who like eating bark and kill the trees by stripping them, as well as by pushing them over. Elephants first appeared in the Ndutu area only five or six years ago. Now they are on the increase throughout the Park and Conservation Area because they are coming in from areas outside where they can be legally hunted. Anyone who has seen the havoc that has been wrought by elephants in Kenya's parks must fear for the future of the Serengeti trees.

Acacias have a hard time of it: when not de-barked or uprooted by elephants, their crowns are plucked by giraffes. The charm of giraffes I find irresistible. The appear so gentle, so unlikely in shape, so decorative in their dappled uniform, with their long eyelashes and queer, loping gait. We saw, that morning, several young ones of assorted sizes. Even the smallest, a few months old, was an exact replica of an adult.

Oxpeckers, or tickbirds, pluck ticks off giraffes, and Hugo has observed them running up the long neck, clinging to the lip and thrusting in their bills to drink the animal's saliva. One giraffe got irritated and tried to dislodge the bird by shaking his head. The bird persisted, and the giraffe kept it at bay by sticking out his long tongue and keeping it pointed at the bird as it flew from one side of the giraffe's mouth to the other, trying to find a way in.

The bush petered out into a strip dotted with whistling thorns, those stunted-looking little trees bearing black galls which are full of ants. Then

Maasai giraffes 'necking' – fighting by swinging their heads at each other. Although usually gentle and apparently a form of play, one giraffe was seen to knock out its sparring partner for fifteen minutes.

came the plains. They are very flat, treeless plains and stretch in all directions to a rounded horizon, such as you find at sea; in fact the whole plain puts one in mind of a vast ocean, as a rule a brown one except when rain, almost overnight, turns it green. The illusion is strengthened by the strong and steady wind which nearly always blows. The heat is not oppressive, since most of the savanna lies at an altitude between 4000 and 4500 feet.

Wherever you stand, you can see hills in the distance, lumpy rounded hills, for this is an old, old terrain, eroded for millennia by wind, sun and rain. Here and there a kopje, a cluster of granite rocks worn smooth by erosion sprouts, as it were, from the plain, amid a little oasis of trees – perhaps just one tree – and bush. These kopjes (inselbergs) have been described as the tops of mountain ranges which lie under volcanic ash thinly spread over the vast plain.

The sky is seldom altogether empty of clouds. These African clouds have much more colour in them than European ones: different shades of grey from almost pure white to charcoal, with hints of violet in their folds. When storms come up, their colour is a savage indigo. In another way they differ from their European counterparts: only very seldom do they get between you and the sun. They seem to part, as they drift across it, to let the sunlight through. Their shapes are varied and beautiful. Often they look like whipped cream piled in twists and peaks in the sky.

We came to the wildebeest migration soon after seven o'clock and were driving slowly through the glistening armies for about four hours. These pewter-coloured, white-bearded creatures move in long, close-packed columns at a steady pace and in a constant direction and, when the time comes to graze, spread out as far as the eye can see in every direction – and probably the eye can see for about twenty miles – like great hordes of ants speckling the plain. All the time, when on the move, they emit harsh grunts, something like the sound of frogs, something like that of old men clearing their throats. People have called them ungainly because of their high shoulders and sloping hindquarters, and also clowns because of their long pale faces and white beards, but in fact they move with grace and sometimes playfulness, leaping and cavorting with apparent *joie de vivre*. Their heads go down, their tails go up, they bounce like balls, kicking up eddies of dust. When disturbed they gallop off together and make curious jinking swerves, at the same time lashing their tails. They do this, Hugo said, when they want to see what is going on behind them without stopping to turn their heads. Most antelopes have a wider angle of vision than the wildebeest.

61

There are said to be about two million wildebeests or gnu (*Connochaetes taurinus albojubatus*) on the Serengeti. It seemed as if we saw them all that day. Of course, we saw only a fraction of their numbers; other armies, perhaps bigger still, were on the march across a front of maybe a hundred miles. Their numbers are increasing year by year. Hyenas, lions, cheetahs, wild dogs and leopards do their bit to keep them down, but prey has outstripped predator. Wildebeests are much like smaller, lighter bison, and this is what the plains of North America must have looked like a century and a half ago – minus the Red Indians.

Every year, the wildebeests migrate in search of fresh grazing within an area of perhaps fifteen thousand square miles. When rain falls on the open plains, they come in from the bush lying to the north, west and south to enjoy the short, palatable grasses which spring up immediately after a downpour or two, and which are rich in calcium. If the rains come according to plan (quite often they don't) they start in December and continue, on and off, until May or June.

The wildebeests' term on the short-grass plains is timed to coincide with their calving, so that the calcium will enrich their milk and give a good start in life to the calves. This calving is astonishing: it all takes place within a month. A positive deluge of calves overwhelms the predators. A great many are, of course, taken; hyenas in particular gang up and chew them to bits as soon as they are born; but a great many more survive.

Why, I wondered, don't the predators increase to keep pace with their prey? Most predators are strongly territorial in habit and don't, in normal circumstances, move far away. So there is more than enough to eat when the plains are green, but in the dry season the prey treks away, few living creatures remain and the cupboard of the predator is bare.

The first wildebeest calves arrived on the Serengeti at about the same time as I did, in mid February. That first day, we saw scarcely any; a week later, a lot; by the end of the following week thousands of them were galloping beside their mothers or resting beside them on the grass. Although adults are silvery-grey, the offspring are light brown with black faces, and save for their sloping quarters look rather like Guernsey calves. The speed with which they recover from birth is amazing. One minute a calf, wrapped in a membrane and trailing an umbilical cord, is deposited on the ground. Within five minutes, quite often, and seldom more than six or seven, it is up and away, galloping with its mother on long spindly legs and shaking its head.

The process is fascinating to watch. The mother, who gives birth lying down, immediately gets up, thus rupturing the cord. She may gently nose the infant, but the first one I watched did nothing to help it to its feet.

Wildebeests near the Gol Mountains, where they share the land with the Maasai and their cattle.

Within a few minutes it was struggling to rise. At first it fell over every time, but each try was better than the last until it managed to stagger a few paces before collapsing again. The mother just looked on. It got up again, still tottering; the mother walked away; unsteadily, the calf followed; within the regulation six or seven minutes it was galloping after her to join the herd. Wildebeest calves, scientists have said, gain co-ordination quicker than any other ungulate. This is a necessary condition of survival. Five minutes of helplessness are plenty for any hyena. The mothers try to fight off attackers, but seldom succeed. If rains are normal, the wildebeest armies remain on the short-grass plains until June, and then trek west and south-west where short grasses give way to longer herbage and to bush.

Although they are nomadic, at breeding times especially wildebeest bulls are governed by an instinct that must have been implanted in them long before they took to roaming over the plains. (The theory is that they, like other antelopes, were originally forest dwellers, and adopted habits of migration to exploit new food supplies.) Before they mate, however, bulls must stake out a territory. It may be a very small territory, no more than thirty yards across, and they may hold it very briefly, for perhaps only a few hours, before they move on. But hold a territory they must before they mate.

So, at the time of the rut, which takes place in May or June, there are many fights between bulls, some serious but mostly 'challenge rituals' to warn off others. These displays are constantly repeated with a tossing of horns, a pawing of the ground and a great deal of grunting and roaring. During the rut the bulls have a hectic time, simultaneously chasing off neighbouring males and trying to keep females within their territory. The females just try to keep moving on.

The migration may look chaotic, with animals mixed up together in these ever-moving herds and columns, but a structure underlies it all. Like most animal social structures, it is centred on the family. These armies are organised into nursery herds of related females with their young, averaging in numbers perhaps ten or twelve, and bachelor herds of young bulls who have been expelled when about a year old from their own families. They are ready for the rut at rather more than three years old. Then comes the brief struggle for territory, always at the same stage of the migratory cycle. When the rut is over, bulls revert to the status of bachelors until the territorial struggle is renewed the following year. But even on the open plains when there is no mating, bulls will mark out transient territories and you will often see a sparring match.

After the rut, migration continues north-westwards out of the Park and into the district of a people called the Ikoma. Here there are *shambas* (small farms), scattered villages and bush. It is here that the animals are most vulnerable to human predators. Rain should fall (again, it doesn't always) in the Maasai-Mara district of Kenya in November. Ignoring the international boundary, the wildebeests proceed into Kenya's Park. When rain, if it comes at the right time, has made the ground soggy – a condition liable to give them a form of foot-rot – they turn south-eastwards and trek back to the Serengeti's short-grass plains, which lie partly in the Park and partly in the Conservation Area. Then the cycle starts all over again. There is a separate population of wildebeests in the Ngorongoro Crater, part of which stays within the crater and part of which migrates outside.

Wildebeests are not the only herd animals to migrate. We passed an army of zebras also on the move. In the main, the two species stay apart; but there is a good deal of overlapping on the fringes, and you often find a small party of zebras in among the wildebeests. At water-holes, they will often drink together. Zebras are also organised into family parties, but have a harem structure; the male generally stays with his mares and foals and tries to protect them. He will put himself between them and an attacking predator and defend them with teeth and hooves. Hugo has recorded an occasion when a family of ten galloped back to the rescue of a mother and foal being attacked by wild dogs. The foal was saved.

*Two wildebeest calves search for their mothers having been separated during a lake crossing.*

This account of the migration outlines what should happen, and what does happen in so-called normal seasons, but there is little normality about African rainfall. This mainly depends upon monsoon winds blowing in from the Indian Ocean, and winds are notoriously fickle. There

is a secondary influence centred on Lake Victoria; precipitation increases as you go from east to west, and the lake basin is wetter than the plains.

Failure of the rains is the great scourge of eastern Africa, bringing famine, thirst and death to humans and to countless animals. This has always been so and is likely to get worse, not better, as populations multiply. In Tanzania, humans are multiplying at the rate of about three per cent a year. The wildebeest population has increased fivefold in the last twenty years, but scientists believe that saturation point is still some distance away. The resources sustaining both, the crops and grass and springs and rivers, have not multiplied.

In 1981, the rains were capricious. Enough had fallen early in the year to green up the plains, and the migrants had arrived on time. But little fell in February, and by early March the grass had been eaten down to its roots and the plains were turning brown. So the herds were leaving for their dry-weather refuge in the bush and coarser grasses to the south-west, and what we were seeing was a secondary migration within the main one. As soon as rain fell again to flush those resilient grasses, back would come the herds from the bush.

With all these thousands on the move, it is not surprising that many calves get separated from their mothers. It is distressing to see these small creatures with their soft brown hair and long black faces stand alone on the vast plain and bleat, or run towards every adult they see in the hope, only too often a vain one, of finding their dams. Meanwhile, there are mothers distractedly galloping about looking for their lost children with equal lack of success. Cows have been observed searching for a lost calf for as long as three days.

No cow will accept a calf not her own. How do mother and child recognise each other? By the sound of her voice, I was told, in the case of the calf; but the mother knows her child by her sense of smell. Two million wildebeests, each with a different grunt! It seems incredible. But then, one thinks of people. We can distinguish each of our acquaintances, if we shut our eyes, by the sound of his or her voice. In a hall full of Chinese, I should not be able to tell one voice from another, but each of the Chinese could do so. Nevertheless I still find it extraordinary that each of those frog-like, wheezing grunts should sound a different note, undetectable to a human ear but recognisable by a day-old wildebeest.

Much of the harm is done, as regards this separation, when something disturbs the animals and they move from a steady walk into an agitated gallop. The disturber may be a predator, but it may be, and often is, a tourist bus or someone's Land-Rover. Habitués like Hugo drive carefully and try not to alarm the animals, but drivers of tourist buses are seldom so

Being used to crossing rivers during the great migration, wildebeests often try to cross small lakes, such as Ndutu, apparently unaware that they could go around it.

considerate, and tourists themselves want to get as close as possible. Pilots of low-flying aircraft showing clients the migration are also to blame. Away go the herds, and many calves are left to certain death, either at the jaws of predators or from starvation.

We came on many of these orphans. One ran right up to the Land-Rover, and we tried to unite it with the nearest herd. As soon as the calf caught sight of the column it galloped off, and we saw it running from animal to animal, but none would accept it. They merely moved away. There was nothing more we could do.

One evening we returned to camp to a scene of a tragedy. Large numbers of wildebeests, still trekking westwards, had reached the shores of Lake

Ndutu and, instead of going round – it is not a very large lake – had decided to swim for it. They are accustomed when on the move to wade across rivers and perhaps thought they had merely reached another, wider than the rest. Half of them at least had calved within the last two or three days. The tottery little calves plunged in beside their mothers and soon found themselves out of their depth. Strangely enough they can swim almost from birth and quite a lot reached the other side, crossing about four hundred yards of water.

But the wildebeests had panicked. Some of the cows had turned back half-way, leaving their calves to struggle on; others had got across but the calves meanwhile had drowned or lost their mothers and turned back. We drove down to the lake shore to meet a harrowing sight. On the far side a horde of wildebeests galloped to and fro along the water's edge, lowing and croaking; on our side many little calves, soaked and miserable, ran frantically about seeking their mothers. A great many drowned bodies floated on the surface of the lake. The air was full of noises of distress, and vultures had gathered in hundreds all along the shore, too bloated to fly. One or two, on our approach, lumbered for a few yards trying to take off and succeeded with difficulty; but most just stood there, I suppose waiting until they had digested one meal sufficiently to make room for the next.

Several calves cantered up to the Land-Rover, uttering soft moans. Some were dripping, others had dried and were shivering forlornly. The sky was reddening and soon night with its cold and predators would end their short lives. Some, hearing the calls of lost mothers on the other side, went back into the water, tottered a few steps and stood there, too weak to swim but still drawn towards familiar sounds. Many must have recognised their mothers' calls.

One of the calves came up to the Land-Rover and lay down beside it, shivering and grinding its teeth. I stroked its ears and wondered, stupidly no doubt, whether it could derive even a grain of comfort from this contact with another living being; a being, alas, impotent to help. A rescue operation on a scale to match the disaster would have been quite impossible to mount and is, in any case, against the rules of national parks, where natural processes must take their course. This drowning of calves, so distressing to see, is part of a mechanism for population control essential for the species' survival. In 1973, over three thousand calves died in this lake alone.

A party of wildebeests who had made the crossing galloped along the lake shore calling for their infants. All the mobile calves dashed up to them but there were no reunions. Some of the cows plunged in and swam back to rejoin the main body on the other side. Three wildebeests complete

During a night-time crossing of Lake Ndutu, just below Hugo's camp, a lion roared and panicked the nervous wildebeests causing 400 to drown.

with calves appeared and also plunged into the water. Their calves followed, got part of the way across, fell behind and struggled until the waters engulfed them and we saw them no more.

More wildebeests from the far shore entered the water and started to cross. They swarmed right through a line of flamingos feeding in the shallows. Some of the birds took off, flew around and re-alighted, pink changing to a deep crimson as they spread their wings against a crimson sky. Beauty and tragedy mingled in the slanting evening sunbeams over dark silver water, with spreading trees beyond.

The little calf beside the Land-Rover was still grinding its teeth and huddling against a wheel. We had to move it; it tottered a few steps, collapsed, got up again, staggered a few more paces and again fell down. For the last time, I thought, but as we drove along the shore several prostrate little figures suddenly kicked out their legs, got to their feet, fell again and kicked out in a dying spasm. Next day the shore was littered

With a friendly touching of heads, a lion is about to settle down next to one of the lionesses in his pride.

A male looks on while his lionesses, gently holding tails, relax by the edge of a *korongo*.

Overleaf: In Lake Manyara National Park lions commonly rest on tree branches.

with corpses, but a few calves were still alive, still bleating for their mothers under the vultures' greedy eyes. I recognised from its position on the beach all that was left of the calf I had stroked the evening before – a wisp of skin, some vertebrae and a miniature skull.

Bones, skulls, horns lay all about. Queer protuberances appeared to be growing out of some of the horns which had evidently lain there from previous disasters. They were about an inch long and a dirty white colour with coruscated surfaces, like blunt-ended spikes. These were the pupae of a kind of moth whose grubs feed on the keratin of which the horns are largely made. So nothing is wasted.

Serengeti lions must be the laziest on earth. They lie about on rocks, on open savanna, beside a reed-bed, in attitudes of supple abandonment, gazing over the plain through hard cornelian-like eyes with pin-point pupils: not hostile, not fearful, just indifferent. They can look more utterly relaxed than any animal I know except perhaps their distant domestic cousins. Cheetahs and leopards somehow manage, even when relaxed, to look alert. Lions just don't bother.

Admittedly, I saw them at a time when they could indulge their laziness *à l'outrance*. Food was everywhere, and didn't even need to be hunted. Helpless lost calves were all about. There is not much meat on a newborn wildebeest calf, but enough to be going on with. In lean times, when the herds have gone, they will even catch flamingos, and there is precious little meat on a flamingo. Hugo once watched a pride during the dry season, and out of sixteen cubs, eight died of starvation.

Unlike a wildebeest, a lioness will accept the progeny of other females and suckle them if she is able. The pride is basically a nursery unit, consisting as a rule of several related females and their cubs. The males, who are quite often brothers, generally stay apart from the nursery group but not far off, and claim priority, not always with success, at the kill.

The first pride we came upon consisted of four lionesses lying in a group together, with sixteen cubs of varying ages. They paid no attention to the Land-Rover, even when it approached to within about twenty yards. Now and again one of them turned over to expose its tummy to the sun, paws in the air, the essence of the voluptuous, and seemingly boneless. After a while one of the lionesses got up, strolled over to a rock beside a small *korongo* (a shallow depression, moist and sometimes filled with water), lay down and stared across the plain with unblinking eyes. A butterfly settled on the rock beside her. With her big flat paw she flicked away the butterfly, which flew off in time to avoid getting swatted. It flew around, returned to the rock, she tried again to swat it and failed. They played this game, lion and butterfly, for several minutes.

We settled down to wait in the hope that, when shadows lengthened and antelopes and zebras made their way towards a shallow water-hole nearby, the lionesses would prepare to hunt. We had already seen the pride's males, a pair with fine manes, sunning themselves about a mile away. The cubs were too young to hunt but, when a kill had been made, their mothers would summon them with soft grunts to share the meal.

A lot of Hugo's life is spent waiting for something to happen. I asked him if he ever passed the time by reading. He had at one time done so, he said, but found that he missed too much – not necessarily an action by the animal he was watching, but the movements of other creatures, birds perhaps and insects, all around. So now he just sits, looks, smokes and at intervals drinks a cup of black coffee. Modern safari life has this disadvantage: one sits too much, and has too little exercise.

One of the largest of the cubs, a teenager, aroused my hopes by getting up, detaching himself from the heap and going through the motions of a stalk. His eyes were fixed on a distant group of zebras, much too far away to stalk even by the most experienced of lions. The cub advanced about fifty yards and then sat down, still looking at the zebras. None of the others joined it. As the shadows lengthened, it became apparent that the lions were not going to hunt that evening. Provided that they make a good kill, lions need to hunt only once in about three days, and these looked well-fed.

Hopes were raised on another occasion when Hugo spotted the head of a lioness all but concealed in a reed-bed beside another *korongo*. A file of wildebeests was passing by not thirty yards from where the lioness lay. It looked like a hunting situation. A wildebeest stopped to drink just

Lions can blend perfectly with their surroundings. In one case a family on a picnic did not realise they were being watched by two lions until they were seen on photos taken of the picnic.

One of a pair of crowned cranes which nested below the camp at Lake Ndutu.

opposite the lioness and three feline heads, not one, rose cautiously above the reeds.

The wildebeest raised its head and looked, so far as we could make out, directly at the lionesses. It must have seen them. There was no reaction: no alarm, no fear, no flight. Calmly, the wildebeest turned and walked after the rest of the herd. It must have known that the lionesses, even though positioned for attack, were not hunting. How it knew is a mystery. Hyenas, lions, cheetahs, mingle so closely with their prey on these plains that Thomson's gazelles, zebras and the rest will drink from a pool beside which hyenas are lounging, and graze within a hundred yards or so of a pride of lions. The prey knows when the predator is satisfied or resting. Then suddenly the curtain goes up on the savage drama of chase and kill.

More wildebeests were advancing towards the *korongo*, and with them several lost orphans running to and fro. Also there came several bereft mothers, not the right ones. They passed beside the crouching lionesses; still no reaction; then one of the orphans came bleating back. It ran to and fro beside the *korongo* under the eyes of the watching lionesses. Surely, now, they would stalk and spring. But they did nothing, and the orphan ran forlornly back in the direction it had come from until it became a tiny speck on the plain. We consoled ourselves by watching a pair of handsome crowned cranes with their golden crests stalking about among the reeds, and a charming brown and white marsh owl crouching under a grass tuft only a few yards away. The pasture was starred by a small white petunia-like flower, and decorated by a little yellow daisy.

Bird life is rich round these *korongos*, especially one known as the Hidden Valley – scarcely a valley but a shallow depression unmarked by trees and therefore hard to find. Birds in thousands flocked to this green marshy ground: ducks; big, handsome Egyptian geese; flocks of pure-white terns; stilts picking their way fastidiously about; blacksmith plovers nodding their heads; Caspian plovers in restless coveys, crowned plovers, white and black European storks wading in the marsh, clumsy-looking black open-billed storks – their bills do not close completely and they filter water through the gap to catch small frogs and other aquatic life. Many swallows, including our own familiar European swallow, swooped about, and a bird related to them, though ground-dwelling, the pratincole, was crouching in the grass beside the pans, on the lookout for flying insects. A pallid harrier, a bird which resembles the more familiar Montagu's harrier,

Overleaf: The wildebeest migration heads west at the start of the dry season.

swooped over, and there were kestrels and a distant eagle, perhaps a martial eagle, which will capture monkeys and even small antelopes. Altogether this is an ornithologist's paradise.

A paradise also for hyenas, who love water and were lying half-submerged in pools or basking on the verges. They are cheeky animals. One of them approached with a hunk of meat in its mouth, probably the leg of a wildebeest. It waded a short distance into one of the pools and dropped the meat into the water. Hyenas use these shallow pools as larders. Two marabou storks followed, probing the marsh-grasses with their long sharp bills in search of meat which they evidently knew had been taken from a kill by the hyena. They kept to the verge, and didn't explore the water. The hyena had fooled them. After a good drink it lay down a little way off and dried its spotted fur in the sun. Unhurriedly, the marabous stalked on.

When the sun was low on the horizon we headed back to camp past Two Trees, a landmark as every lone tree, or pair of trees, must be in this treeless landscape. They were loaded down with vultures as with heavy fruit, and they also sheltered owls. Competition for roosting-space must be acute. Here, still moving, still grunting, still whisking their tails, were the wildebeests again, golden-bearded as the setting sun shone through their tufts of hair. The dust which floated over them was golden too and their silhouettes were black against a crimson evening sky. The scene was dramatic, also sad as evenings are when they are beautiful. Another African day is dying.

The Land-Rover jolted over tussocks and jinked to dodge holes as Hugo sought the right position to photograph the scene. The wildebeests had broken into a gallop. As the light faded, a solitary one came galloping back, sharp against the skyline, bucking the traffic in search of a lost calf.

We drove back in the dark for most of the way. The Land-Rover has excellent headlights, but every time an animal or group of animals loomed ahead Hugo switched them off so as not to cause alarm. When startled by beams of light zebras especially are liable to panic, wildebeests mislay their young, even tommies (Thomson's gazelles) lose their heads and dikdiks take flight and snap a match-stick leg. The road was pitted with holes, and we had to churn through a swamp too deep for anything but a Land-Rover and almost too deep for that. But we evaded disaster; the camp was welcoming and Sirili had hot soup ready and a tasty cottage pie.

# Problems at Seronera

You cannot be long in any East African park without hearing talk of the tragic fate of the black rhino. The general belief used to be that the main use of the rhino's horn was as a (supposed) aphrodisiac in India and the East generally, but this is not so. A study of the trade in rhino horn carried out on behalf of the World Wildlife Fund by Dr Esmond and Chryssee Bradley Martin demonstrated that there are two principal uses for the horn: in traditional Chinese medicine, practised in many Eastern countries, and to be carved into decorative handles for the daggers which are a status symbol in parts of Arabia, notably the Yemen. In a sense the rhino has been a victim of the massive increase in the price of oil that occurred in the 1970s, and which enabled so many young men to acquire these fashionable daggers.

The Martins' account of their investigation, published as *Run Rhino Run*, makes sad reading. In 1966, seventy rhinos inhabited Olduvai Gorge in the Serengeti: by 1979 there were none. In the same period, nearly

An increasingly rare sight: a black rhino (with pointed lip) pulls a face on smelling the track of another. Tickbirds or oxpeckers remove parasites and flies from these beasts' hides and warn their short-sighted hosts of danger.

three-quarters of the rhinos in the Ngorongoro Conservation Area were wiped out. The largest concentration of rhinos in the world used to be in Kenya's Tsavo National Park, and as recently as 1969 Dr John Goddard, who studied the species for three years, estimated their numbers at between 6000 and 9000. The latest estimate I have heard is less than two hundred survivors. This population crash was not solely the handiwork of man, whether poachers, farmers or hunters. Following a devastating drought in the early 1970s the bones of rhinos who had starved to death littered huge areas of withered bush surrounding dried-up water-holes and river-beds. Experts reckon that over Africa as a whole, the rhino population has been halved in just one decade, 1970 to 1980: and the slaughter goes on. 'If the rangers do not improve their technique, then rhinoceroses are doomed in northern Tanzania,' the Martins conclude. Things are little better, if at all, in the rest of the country and in Kenya.

No one knows just how many survive on the Serengeti but the present estimate is 'fewer than fifty'. It may be a lot fewer. I saw exactly one rhino

Speared by a Maasai before dawn (while tickbirds sleep) in Lake Manyara National Park this picture was taken as the rhino died. Unfortunately, a lucrative illegal market for their horns ensures the extinction of this extraordinary creature.

during my whole stay, a female all alone on the shadeless plain, very conspicuous and pathetically vulnerable. You could not have found an easier target. Rhino calves stay with their mothers until they calve again, by which time they are quite large sub adults. This rhino had no calf with her. She heaved herself to her feet when we approached and lumbered away. What was her life expectancy? And, with so few individuals in so great a region, what were her chances of finding a mate?

Rangers on the Serengeti and in the Ngorongoro Crater are charged with giving special protection to the rhinos. One day, exploring some kopjes, we came upon two or three rondavels half-hidden in dense bush. This was a ranger outpost, a longish way from Seronera, the Park's headquarters, and served by no road. These rangers do indeed live out in the blue and their lives can be perilous as well as lonely.

Poachers fall into two broad categories: those who are after meat, and those after trophies. The former are generally tribesmen living near the borders of the Park who hunt with poisoned arrows, traps and snares, and concealed pits. Poachers belonging to the latter group are much more dangerous and sophisticated. They have fast vehicles and modern automatic weapons and generally operate in well-organised gangs. Some are Somalis who, in the 1970s, moved into the business in a big way. By contrast the Park rangers have antiquated weapons and are chronically short of vehicles, sometimes of petrol, and of such necessary aids as adequate radios, light aircraft and so on. They are not even well paid. Poachers make enormous profits and it is not surprising that rangers are sometimes tempted to look the other way or to turn poacher themselves. A good pair of rhino horns may be worth considerably more than a year's pay to a ranger.

To make a photographic record of an anti-poaching sortie, Hugo accompanied a team searching for a gang in the western section of the Park. The poachers were after meat, to be sold at a handsome profit in villages across the boundary. The rangers had found a line of pits, each one about six feet long and six or seven feet deep, linked by thorn-bush fences. The animals would walk along these fences seeking a gap, and fall into pits cleverly concealed by branches. There, if the poachers didn't find them next morning, they would die of injuries and starvation, and terror. The poachers, armed with spears and poisoned arrows, had a hide somewhere in the bush where they butchered and dried the meat. The task of the team was to find it and arrest as many poachers as they could.

The squad consisted of thirteen men with two vehicles. First, they came across a cache of meat cut into narrow strips to dry in the sun, together

with piles of hides. There was a hut nearby, with a fire still burning under a cooking pot. But the poachers had fled. Next day they spotted three men, gave chase and caught two of them, who had bows and poisoned arrows.

The main gang eluded them until, two days later, they apprehended an old man who led them towards the poachers' hide which was hidden in thick bush. The rangers were posted in a semi-circle in the bush while their leader, the park warden, walking behind the hand-cuffed old man and, followed by Hugo with his camera, advanced upon the hide, hoping that the main body of the poachers would flee into the arms of the rangers. Although the plan partially miscarried, four poachers were captured, three of them by a ranger who was such a fast runner that Hugo thought he would satisfy Olympic standards. This man overtook each of the three fleeing poachers in turn, handing them over to colleagues to hold while he pursued the others. When the chase was over Hugo found him

A member of the Serengeti's anti-poaching team uncovers a trap: part of a long line of pits covered by twigs and grass, the intermediate spaces being blocked by a wall of cut-off vegetation.

Hugo with two friends, spent six hours freeing this wildebeest. The poachers usually inspect their traps at dawn and kill such animals with poisoned arrows.

84

standing in the hide in the midst of drying meat, and sobbing. He was sobbing because so many of the gang had got away. There were about forty of them, armed with poisoned arrows but luckily without rifles. Rangers, as I have said, do sometimes yield to the temptation to take bribes and to kill the animals they are paid to protect: but as this incident shows, some at least among them are truly dedicated men prepared to risk their lives to catch the poachers.

The various ungulates of the Serengeti are present in such enormous numbers that, so long as the poachers are kept on the defensive, and humans with their cattle continue to be kept out of the Park, their survival is not threatened. It is the trophy-bearers that are disappearing fast. A few years ago, Hugo photographed several leopards living and breeding in the Seronera region, but they have now disappeared. Although we searched in likely places, we found no traces anywhere. But we did see cheetahs, the solitary rhino, and a high concentration of elephants.

Poachers cannot be blamed for the plight of the wild dogs. In 1968, twelve packs hunted over this part of the Serengeti. Now there are three at most. Every day, Hugo searched the plains with his binoculars, hoping to sight one of the packs he had known – the Plains pack, the Pimpernel pack, the Nettle pack, the Genghis pack above all. Not a sign did he see.

The introduction of distemper is almost certainly to blame for this population crash. No one is allowed to keep domestic dogs in the Park, but in the Conservation Area the Maasai sometimes do so and, for others, exceptions have been made. Hugo told me of a pair of scientists who had camped near him in the Gol Mountains to study wild dogs – the Genghis pack in fact – and had parked within twenty or thirty yards of the den. He was horrified to find that they had with them a sick dog which they had been brushing, and had thrown the loose hair on to the ground.

Even when recalling the incident, Hugo's eyes kindled with anger. 'And they were scientists!' The miscreants were obliged to pick up every hair, and burn the lot. Even if their dog did not have distemper, it had some disease. It was almost certainly distemper that killed off most of the Kühme pack while Hugo had it under observation. The survivors suffered from bouts of shivering, rather like fits. An official estimate made in 1981 put the number of wild dogs on the Serengeti – there are now none in the Ngorongoro Crater – at two hundred. Even that is probably too high.

At Seronera, you come suddenly in touch with the outside world again. The tourist lodge, run by an agency of the Tanzanian government, has been designed so that it incorporates part of a kopje. A big, smooth grey boulder, forming one wall of the lounge, towers above you as you sit over your drinks, and other boulders rear up from courtyards, with staircases and steps linking and encircling them. Rare indeed are attempts to build upon the bones of Africa instead of littering it with ugly alien materials that have no relation to the landscape. It is a most attractive, well-kept lodge, unfortunately suffering at the moment from Tanzania's chronic ailment, a shortage of necessities owing to a lack of foreign exchange. House-keeping is difficult at the best of times, for all supplies must come from Arusha, over two hundred miles away. When we visited the lodge there was no water – the pump had broken down and no spare parts were to be had – and there was also no petrol and no bread.

Nearby is the Seronera Research Institute, which throughout the 1960s harboured a team of ecologists and students of animal behaviour drawn from a number of countries and probably unmatched in Africa for its expertise. In its heyday, about fifteen scientists were engaged on different

Taken by surprise near his hide-out, a poacher with bow and poisoned arrows drops to the ground and surrenders to a member of the anti-poaching unit. The meat of buffalo, zebra and wildebeest is drying in the background.

projects, ranging from research into grassland, forest and woodland ecology and soils, to the ecology of wildebeests, zebras, elephants, buffalos, giraffes, lions, vultures and Thomson's gazelles. The qualifications of the scientists, all of whom were funded by overseas foundations and universities, were impressive; much of the work they turned out and published was outstanding. Dr George Schaller used Seronera as a base when he made his classic studies of the lion, on which all subsequent studies of this great predator have largely been based. So did Dr Hans Kruuk, already mentioned. Dr Iain Douglas-Hamilton's five year study of elephants was no less seminal, and the names of Dr Hugh Lamprey the director, Dr Sinclair from Texas who worked on buffalos, and Dr Philip Glover, who knew as much as any living man about the grasslands of Africa, were internationally known in scientific circles. The Institute was part of the set-up of the Tanzanian National Parks whose then director, Dr John Owen, brought it into being and was its most staunch champion.

All this was built up after Tanzania obtained its independence in 1961. In 1972 there came a change of policy, and a new director. In a sense the SRI was a victim of its own success. Too much effort and expertise was being concentrated in one institute while others had their own rightful claims to limited resources. Too much of the long-range research had little relevance to the immediate needs of an impoverished developing country. All the scientists were expatriates, not Tanzanians. Such must have been among the considerations that influenced government decisions. At any rate the SRI, while it was not closed down, shed its expatriate scientists, assumed a Tanzanian director, and reduced its scope. No doubt all this was inevitable, but I found it sad to see the Institute a shadow of its former self, and especially the library which had built up such a fine collection of books and journals. But the houses formerly occupied by the scientists are still here, pleasant little bungalows sheltering in the lee of kopjes inhabited by colonies of rock hyraxes. Here also are the headquarter offices of the Park. Seronera has become quite a village – too big, people say. Modern ideas favour siting offices and lodges on the perimeters of parks, so as not to pollute their natural wildness.

# Into the Gol Mountains

The Gol Mountains are not so much mountains as a long, broken line of hills that rise and fall in folds and, as you go eastwards, open into a plain which ends in an escarpment subsiding into the hot, dry wilderness around Lake Natron. Out of this wasteland the Maasai's Mountain of God, Ol Donyo Lengai, rises to over 9000 feet above sea level.

The Gol Mountains lie outside the Park but within the Conservation Area, so Maasai flocks and herds are pastured there when rain falls, but there is no permanent water and in the dry season they are driven elsewhere. Animals that can survive long periods without water, or travel long distances to seek it, eke out a living – a few oryx and Grant's gazelles, an occasional cheetah, some jackals and hyenas and small creatures like mongooses and the ubiquitous lizards that dart into holes as you approach. And of course raptors, who draw moisture from the blood and tissues of their prey. Armies of Ruppel's griffon vultures nest in cliffs on the eastern escarpment – over one thousand nests have been counted there. Each female lays a single egg.

A cheetah approaches a Thomson's gazelle fawn lying motionless and almost invisible in the grass.

Strong winds sweep the Gols and to camp there you must find a relatively sheltered spot. Hugo knew of one such – the spot where the dormouse had nested in the fork of a tree. Umbrella thorns provided shade, a shoulder of the hills the necessary shelter. 'Hugo always had a flair for setting up camp,' Vanne Goodall, Jane's mother, has written. 'In a remarkably short time, and sometimes with the minimum amount of help, the tents would be up, a fire lit, supper cooked, and Hugo, relaxed and at ease, a cigarette glowing in the starlight, would sometimes begin telling stories. He is an excellent raconteur, witty and concise and never malicious.'

With us in the camp were Steve and Sharon, who had turned up at Ndutu in their Land-Cruiser and greeted Hugo as an old friend. Steve is a pilot flying for TWA who has made the East African game parks his second home. He saves up his leave and, when enough has accumulated, flies to Nairobi, picks up the Land-Cruiser, rendezvous with Sharon, and together, with cameras at the ready, they head for their favourite camping grounds. So we shared with them the camp under the trees, a churn of water and, after nightfall, Sirili's skill with a cooking pot over an open fire.

'Perhaps,' Hugo said next morning, 'we shall see the Genghis pack today.' He was always looking for it. The dogs he knew and filmed ten years ago must all be dead by now, but some of the pups could still be living – just. When he last saw the pack, it had dwindled to five members.

We drove down gently undulating foothills and before long came to a cheetah lying under a tree. It was a very placid cheetah. We edged to within ten or twelve yards and it didn't even turn its head to look at us, but lay with its broad cat-face – the black stripe down its cheek like the track of a tear-drop – gazing across the plain towards a group of wildebeests in the distance. Probably it was a female digesting an overnight kill.

Rain had fallen, so hill and plain were misted over with a spring-like green, and wild flowers showed among the creeping grasses. Thomson's gazelles with slim legs as nimble as a ballet dancer's were everywhere, their white and tan bodies with a black stripe shining in the sunlight and their tails always a-wag. Sometimes, when a group of tommies took flight at our approach, a single one would stand its ground instead of going off with the rest, fixing its gaze upon us and wagging its tail. Probably it was a doe whose fawn was concealed nearby under a clump of dwarf sodom apples. Tommy fawns seem to know by instinct, almost from the moment of birth, that on the approach of danger they must flatten out and lie absolutely still instead of following their mothers. How do they know this? How does the mother warn them and subsequently give the all clear?

A puzzling event: normally friendly, these two Grant's gazelles appeared to engage in a serious fight but subsequently departed together, side by side.

The little fawns blend so well into the landscape that you could all but tread on one before seeing it. The mother never wanders far away. Another puzzle. For some reason not yet scientifically explained, they do appear to be in the process of losing their horns. You see some with horns that look half-grown or out of shape, and some with no horns at all. In lesser numbers were their larger cousins the Grant's gazelles, similar except for longer horns and no black stripe.

The odd habit of 'stotting' which both of these species follow is another puzzle. As you approach a gazelle, it will often rise into the air with all four legs as stiff as pokers, just as if it were on springs. It bounces. Why? One theory is that the stotting animal, by drawing the attention of predators to itself, may save the lives of others in the herd. This has been quoted as an example of altruistic behaviour, like that of bees that perish when they sting. Professor Dawkins in *The Selfish Gene* has discredited the whole idea of altruism among animals by arguing that it is merely a mechanism for preserving genes: all creatures, great and small, ourselves included, are in his view so many devices used by genes to ensure their own survival.

Hugo has a different theory about stotting. He thinks that it uses up less energy than running at full tilt. So a hunted animal, or one that senses danger, may resort to stotting for a while to conserve energy for the final sprint. Once he followed a pack of wild dogs in pursuit of a Grant's gazelle for three and a half miles. Dogs and quarry ran at a fairly steady thirty miles an hour and the gazelle was stotting for most of the way. Once or twice the dogs almost caught up with it and it put on an extra burst of speed at a normal gallop. The dogs never did catch it; after three and a half miles they gave up, although they were not themselves exhausted. Possibly they recognised that the gazelle was too healthy for them.

We halted by a kopje with a small pool at its foot. Most of these ancient rocks catch and hold water, thus enabling seeds of trees and bushes to germinate and take root. Ancient indeed are these pre-Cambrian rocks,

A desert rose flowering in the arid region of Ol Donyo Lengai volcano and the Great Rift Valley, on the eastern edge of the Serengeti.

among the oldest on earth, formed millions of years before life began. Much later, in what are by geological standards almost modern times, the great volcanoes to the east and south-east – Ngorongoro, Lemagrut, Nainokanoka, Loolmalasin, Meru and, mightiest of them all, Kilimanjaro – flung forth ash and lava to form the powdery, friable soil under our feet. This, we are told, occurred between two and five million years ago, towards the end of the period known to geologists as the Tertiary.

By this time early hominids, precursors of man, had emerged from among the primates, and eastern Africa is believed, in the present state of our knowledge, to have been the region of our planet where that emergence took place. Those proto-men must have been the first to witness the tremendous eruptions that flung the fiery contents of the earth's bowels far and wide over the world they knew and to hear the roars and rumbles and explosions that must have accompanied those earth-cracking upheavals. Snow-capped Kilimanjaro, over 19,000 feet high, is only a weathered cone remaining when the volcano was spent.

It is easy to imagine how terrifying must have been those mighty explosions, how deep the impression left on primitive minds, and how the fury of offended gods might have been postulated to explain them. Perhaps it was here on these East African savannas that the idea was born of hell and damnation, of the wrath of God ordaining that men might burn forever in fires as terrifying and unquenchable as those they had seen with their own eyes. That this wrath should be visited on sinners would not have been so great a step in the evolution of a code of morals. So perhaps it was here, within sight of volcanoes now extinct, that the thunderbolts of Jove, the wrath of Jehovah and the concept of hell and damnation had their origins.

Although they have calmed down, these volcanic eruptions and the cracking and faulting of the earth's surface are still going on. Ol Donyo Lengai is periodically very active, and the floor of the Rift Valley is sinking at the rate of about one inch every twenty-five years. This does not sound much, but if it continues at the same rate for a million years, quite a short period on the geological time-scale, the valley's floor will end up more than 3300 feet lower than at present.

On the foot-rock of the kopje, two parallel rows of small depressions had been chiselled out of the stone. Here on this rock men of the past, perhaps hunters dwelling beside the pool, played the ancient game of *bau*. The word means board, because today the game is generally played on a board with little hollows into which the players drop beans. It may also be played on the ground after scooping out the necessary number of small pits. It is a very difficult game for most Europeans to master. The object is to capture all your opponents' beans, or no doubt pebbles when this stone 'board' was used.

Who were the players, how long ago the game? Some believe the players were Maasai, the time therefore recent. But although the Maasai do play *bau* and very likely use these hollows today, it seems more probable that they were made by others long before the Maasai came. There is evidence all around of Stone Age man: chips of arrow-heads, hand-axes, scrapers and other tools. Also here and there you come upon stones and boulders grouped in rough circles, possibly the remains of walls protecting their primitive dwellings. Their makers have been called the Stone Bowl people. As well as bowls they also used pestles and mortars, stone knives and scrapers, had crude pottery and ornamented themselves with stone beads. They inhabited these plains until about two thousand years ago.

Here beside the kopje a little of the thrill experienced by those who dig up the past stirs the mind. You stand on ground where primitive men have been before you, dressing their kills, cupping their hands – hairy? blunt-fingered? – to drink from this pool, played their game and wrapped themselves in pelts of animals to sleep in fear, no doubt, of prowling lions and hyenas. And long before that, going back not thousands but millions of years, ancestors of modern man, hominids of species that have left only their bones, may have paused here to drink from this pool. You and I are the latest of a long, long succession of men or man-like creatures. What relics shall we leave? Tin cans, bottles, rusty bits of metal from broken-down motor cars?

A sky-blue flower with a pea-shaped face blooms half concealed in a spiky bush. It has a curious habit: in the early morning you may see a drop of moisture within, but in the heat of the day the flower retracts into this moisture, to open up again in the cool of the evening. Everything that grows must be, so far as possible, proof against heat and drought, against fire, against all forms of attack. You have to look closely to find wild flowers, for many lie low in grass and bush, short-stemmed and inconspicuous. A creeping, tough-stemmed plant with a tiny heather-coloured flower spreads its heath-like claws, and stains the ground in patches as if blood had dried there in the sun.

Everything has thorns, and even insects imitate them. Hanging in one of the bushes was a cluster of pale-coloured thorns about an inch long, tightly cemented together, and all pointing downwards. This was the structure made by a moth whose caterpillar was pupating safely inside the little palisade. Bees, moths, flies, praying mantis, and many other kinds of insect dwelt in the spiky thicket surrounding the pool.

But the pool was shrinking and, if no more rain came, would soon become a pan of dried and cracked mud. Terrapins and frogs inhabit nearly all these kopje ponds. What will become of them? When the time comes they will burrow down into the dried mud and survive. We watched a terrapin basking on a rock, and a companion in the water with its head above the surface, perhaps looking for a meal. Surprisingly, these terrapins catch birds. We waited for some time by the pool in case a bird would come to drink within the orbit of the terrapins, and within range of the camera. Ducks were swimming about, but they were too big for the terrapins. Sand-grouse are favoured victims. They come in flocks to drink late and early in the day, and possess a marvellous mechanism for carrying water to their chicks which lie hidden in the grass some way from the pool. Birds' feathers are so designed as to repel moisture, which runs off the feathers' oily surface. But sand-grouse cocks carry on their undersides fluffy feathers specially designed to hold water. After a quick bath they take off for their nests, where the chicks plunge their bills into their fathers' feathers and drink their fill. Flights of Caspian plovers were wheeling, settling, and taking off again in restless motion. The coveys swept like billowing silver waves across the plain. They are migrants, and were perhaps gathering for their long northward journey. But no sand-grouse came.

People, like wild animals, are drawn towards kopjes because life is there, everything from lions to lizards. As we approached another cluster of boulders we saw that another Land-Rover had been drawn towards it. When a moving spray of dust betrays a distant vehicle in this empty land-scape you almost get a sense of being jostled, like the old voortrekkers who moved on when they could see the smoke of another man's fire.

The occupants of this Land-Rover had stopped to focus their binoculars on three lions who were basking upon the rocks. Hugo greeted the human pair as old friends. They were Craig and Anne, newly arrived from Chicago to continue a study of lions. Both had begun their scientific careers among the chimpanzees and baboons of Gombe Stream.

The lions on the kopje had them guessing. 'I can't make out who they are,' Craig observed, sounding a trifle aggrieved. There was a pride with cubs not far away, he said, but this trio, all males, did not seem to be associated with it and he had no record of them in his charts.

Craig explained how he identified his lions. Every lion has three or four, generally four, rows of whiskers sprouting from its upper lip. Just as there are no zebras with stripes arranged in exactly the same pattern, and no two humans with exactly the same fingerprints, so no two lions sprout

Overleaf: Throughout the Serengeti rock 'islands' called kopjes provide shelter and shade, and sometimes pools of water, for numerous species, including lions.

95

their whiskers in exactly the same number and pattern. Each of Craig's 'study lions' had a page devoted to it in his note-book, a photograph of its face on one side and a drawing of its whisker-pattern on the other. This trio was no doubt nomadic. Male lions leave the pride when puberty approaches and become nomads until they can join another pride, perhaps one whose paterfamilias is losing his vigour and can be driven out. Often, in this nomadic phase, they team up with other bachelors, especially with their own brothers.

The object of this behaviour pattern is to avoid in-breeding. Wild dogs and chimpanzees achieve the same end but by a different means. In the case of wild dogs, young females are turned out by the dominant female. The young bitches then try to join a pack that has lost its dominant female, or they try to displace her, or instead look for an all male pack.

We enjoyed a morning coffee at the foot of the kopje, the lions stretched out on rocks above us staring indifferently with their hard yellow eyes across the plain. Agama lizards, brightly coloured with blue heads and orange backs, scuttled about, tolerated by the lions who will sometimes let them catch flies on their pelts and even faces. I thought what a satisfactory life these young zoologists have, living for part of the year in Land-Rovers in the open and following their study animals of whatever species. (Some are known by the name of their animal; a 'jackal girl' was camped near Ndutu, we encountered a 'serval girl' on the road, and a 'wildebeest man' was based on Seronera; Hugo is sometimes called by Africans the 'wild dog man' or 'picture man'.) The other part of the year they spend at some university writing up notes and working for their higher degrees. So they can enjoy the pleasures of civilisation while looking forward to escaping from them to the freedom of the wild. Wisely, these young scientists tend to pair with each other and then share the excitements, rigours and, no doubt at times frustrations of their work. If they are clever they can regulate their lives, as do their study animals, according to the seasons. 'There was a blizzard blowing when we left Chicago,' Craig said, sipping his coffee in the sunshine beside the kopje.

At sunset we drove back to camp and came once more upon wildebeests trekking away from the hills. Long columns moved through a nimbus of golden dust against a blood-red sky. Zebras, tommies, Grants, were also on the move. Sunlight slanted in long bands of alternate light and shadow across the rolling foothills of the Gols. 'And when the light in lances Across the mead was laid': Houseman's lines come to mind, although this was no mead – steppe, savanna, range would fit better. Hugo's camera

Lion cubs are usually born away from the pride, hidden until they are four weeks old.

clicked almost frenziedly from beneath a thorn-tree whose arching boughs, black against a flaming sky, framed the picture.

Sirili fried steaks on an open fire and we ate under the stars and by the light of a lamp run off the battery of the car. Steve and Sharon had followed a cheetah, photographed lions, witnessed the birth of a tommy fawn. The whooping of hyenas sounded from the darkness, a zebra's bray, shrilling crickets, an owl's call, many other night sounds came to our ears. Presently a half-moon threw light enough for us to pick our way over boulders to our little tents under the trees.

Before sunrise next morning we headed east over a rocky saddle in the hills and down into a valley beyond. The ground here was rough and tussocky and scored like a side of pork by many narrow tracks made by game and by cattle, into and over which the Land-Rover bumped and bounced like a jumping bean. A herd of elands, antelopes considerably larger than the native Zebu cattle, reacted very differently to our approach from the wildebeests and zebras of the plains. They fled. Although the Maasai do not, in theory, eat the meat of wild animals, and in any case are not allowed to kill them, I was told that they do, in fact, hunt elands, whose flesh makes excellent eating. They don't, however, hunt from Land-Rovers, and I do not know why the elands have not got used to these vehicles, as most other animals have.

The hills beyond the valley are very different from the Gols' western slopes: rocky, precipitous, in places sheer cliff. Down by their foot is a huge rock, standing on its own, named Apis Rock, although its Maasai name is Nasera. It was named Apis not after the sacred bull of Memphis but as a kind of pun. When Louis Leakey and a colleague camped here to investigate a rock shelter, they drew their drinking water from a pool at its base until they realised that baboons dwelling in large numbers on the rock habitually pissed into the pool. From crevices in this mighty boulder, fig trees dangle long roots towards the ground, and klipspringers live here all year round.

Nearby is a cave littered with bones, including the remains of a human skeleton. A Maasai warrior from across the Kenyan border, one of a cattle-raiding gang, had lagged behind his companions. Tanzanian warriors found and speared him, and threw his body into the cave.

Maasai raids continue despite efforts by the police to stop them. Distances are great, warriors fleet and hills and valleys offer hiding places for raided cattle. Often members of a raiding party will start a grass fire behind them to cover their tracks – a problem for the Conservation Area's wardens who try to keep destructive grass fires under control.

Klipspringers, one of the pigmy antelopes, live on rocks where they have found protection from most potential predators by their ability to climb almost sheer cliffs.

Cattle belonging to tribes who live outside the boundaries of Maasailand are fair game. But sometimes the young men of these much despised (by Maasai) tribes answer back with poisoned arrows. A recent battle ended with fifteen Maasai dead and several hundred of their cattle driven away. A sympathiser, talking to a Maasai elder who had lost most of his cattle, enquired: 'How long will it take you to make good your losses?' The reply was: 'About a week.'

Beyond Apis Rock, we met some donkeys with small boys herding them, and came to an old *manyatta* under the lee of a heap of rocks. Most people use the word *manyatta* to describe these enclosures encircled by a thorn-bush fence, and containing the low mud huts, shaped rather like fat

A Maasai *engang*, more commonly but incorrectly called a *manyatta*, at the base of Nasera Rock on the Serengeti. Belonging to a single Maasai family, this *engang* is only temporary because here water is available only in the dry season.

sausages, of a Maasai family. But the word, I am told, is so used incorrectly. A *manyatta* is built especially for the warriors, their mothers and their (uncircumcised) girl-friends. Formerly the mothers built forty-nine huts, this being a lucky number, and here the warriors and their attendant females lived in luxury by Maasai standards: that is, well supplied with milk and meat, and setting forth at intervals on cattle raids and lion hunts. Life in the *manyatta* cemented together the warrior class of each clan and made them, in the words of the Maasai writer and ecologist Tepilit Ole Saitoti, 'more alert, proud and competitive.' It also sorted out a dominance hierarchy, which might be compared with that of wild dogs and other species, enabling qualities of leadership to emerge. Physical courage was the quality most admired. Nowadays *manyatta* life, as it then was, has virtually faded away. The correct name for the family dwellings you see tucked into hillsides all along the Gol Mountains is *engang*. The huts lie up against an encircling fence, the centre of the enclosure is kept clear for livestock, and at night the entrance is closed by thorn-bushes to keep out lions, hyenas and, possibly, human raiders.

When a Maasai family, or group of related families, moves on, they desert the current *engang* and often re-occupy an old one after repairing it. That is what the women we came upon were doing. It is women's work to build and repair. These were young married women clad in leather aprons, with shaven heads, and the traditional wide Maasai collar made of tightly threaded, brightly coloured beads. These collars are handed down from mother to daughter and new ones are also made.

All that was left of the old huts were branches cut from the bush and bent to make a cage. Cow-dung is used as plaster inside and out. Hides, which donkeys carry when a move takes place, are used on the outside to protect against weather. A fire is lit inside each hut which each wife shares with sheep and goats as well as with her children. The fug is formidable, and helps to discourage fleas and jiggas (a kind of flea that burrows under toenails, where its eggs subsequently hatch).

But nothing can discourage flies. Where there are Maasai there are cattle; and where there are cattle there are flies. I suppose that, if the people hadn't learned to live with them, life would be impossible. However much we may respect a people who, since the first white explorers entered their country, have been admired for their bravery, independence, indifference to the materialist values of Western society, and for their bodily beauty, those of us conditioned to dislike flies could not live as a Maasai. Flies cling to their persons, crawl about their faces and into ears and eyes and nose; children's eyes are often encrusted by the insects, yet the children do not seem to mind. We watched a tall, slender

Using strips of bark, a Maasai woman builds a hut in the *engang* near Nasera Rock.

young woman tying in the sticks of an old hut's framework, and counted five flies clinging to the inside of her lip.

Taking photographs of Maasai people is tricky. Because, unlike most other Africans, they value their traditional ways and retain their clothing, hair-styles, ornaments and body decoration, and because they are a handsome people, they have for long been magnets for tourist cameras. For years, when tourists were fewer, the Maasai put up with this – I have often wondered how we should react if foreigners continually pointed cameras at us while we were digging in our gardens, doing our shopping or parking our cars – but soon tumbled to the commercial opportunities. Now the would-be photographer who fails to pay a substantial sum in advance courts disaster. There was a recent case where a man who had included Maasai cattle in a view of the landscape was confronted by a posse of furious warriors shaking their spears and demanding, with menaces, one hundred pounds. In the days of exploration, travellers in

Maasai with their belongings, saved from a bushfire which destroyed some of their huts.

Maasailand complained, often bitterly, about extortionate demands for *hongo*, which they had to pay in beads and calico before they were allowed to proceed. This is an ancient African custom and, with photographers replacing explorers, it still goes on.

Three elders stalked up to us with their long, loping stride, shook hands and enquired our business. Only one of them could speak Swahili and he only a few words. Hugo's cameras spoke for him, negotiations were opened, the elders accepted five pounds and the deal was amicably settled. No tourist buses come into these hills and so the sum was moderate and the bargaining brief.

The elders wore the traditional Maasai toga hanging to just below the knees. Formerly it would have been made of cotton material dyed with red ochre, but nowadays a light woollen material, gaily patterned, is preferred. The elder who conducted negotiations wore a toga with large red checks. He had the thin, aquiline face and prominent Roman nose common among his people, and said by some to resemble the bone structure of the ancient Egyptians. (The Maasai are believed to have entered eastern Africa from the Nile valley.) The women were at first camera-shy but, on the order of the elders, went on with their work.

A warrior strode up, exchanged greetings with the elders, planted his spear in the ground and assumed the characteristic stork-like posture of standing on one leg and pressing the instep of the other foot against the side of his knee. He looked thoroughly traditional, his hair done in a mass of tight little pigtails hanging down over his shoulders, his body anointed with red ochre and fat. Although some of his customs have had to go or be modified, his way of life and his beliefs have not changed much since his ancestors came into occupation of these hills and plains two or three centuries ago. But cattle raids now are fewer and more furtive, taxes must be paid, lion hunts are clandestine instead of great occasions when warriors proved their worth, and the government expects the men to wear shorts under their togas when they go anywhere near a town. A good deal of drama and excitement must have gone out of life. But there is something left: this young man can still enjoy the prestige of the warrior, the freedom of lovemaking with young girls, the dances and the camaraderie of his age-set. He can still keep alive in his heart an unshaken belief in the superiority of his own kind above all others. He would not be at home in our would-be egalitarian and non-racist society. And as for women's lib . . .

Our Land-Rover was of much less interest to him than a cow would have been, so he at first ignored it, but after a while strolled over to admire

Entrusted with protecting the tribe from man and animals, the *morani* (Maasai warriors) sometimes raid cattle from neighbouring tribes, believing they are taking what is rightfully theirs since 'God gave all cattle to the Maasai'.

his reflection in the wing mirror. The elders followed suit and perched on the bonnet, dangling their legs over the radiator. The Maasai display a natural arrogance, a quality which most Europeans deeply resent when their own compatriots exhibit it, but admire with equal intensity when it is displayed by members of an African race.

More women now appeared, including several unmarried girls who came up to talk to the warrior. He preened himself: there is no other word. There is a proverb among the Kikuyu, whom the Maasai deeply despise: 'A young man is a piece of God'. I am sure this young man would have agreed. So greatly does his god Lengai favour his chosen people that he gave them all the cattle in the world when the world was created, so that when the warriors embark upon a raid, they are merely bent on reclaiming cattle that barbarians have wrongfully stolen.

These Maasai, away in the Gol Mountains, have as yet suffered little from the pressures of the modern world. But what does the future hold for this young man, and for the girls with their wide beaded collars and lithe movements and fly-bespeckled faces who clustered around him? The wind of change is blowing away a lot besides the departed white *bwanas*.

Already there are Maasai who have forsaken the old ways. Maasai ministers are in the government in Dar es Salaam. Ole Konchella, director of the National Parks, is a Maasai and so is Ole Saibull, director of the Conservation Area. There are Maasai civil servants, teachers, doctors, professionals of various kinds. A few of these have renounced the fleshpots and power levers of the cities, thrown off their Western suits and come back to herd their cattle in togas and sandalled feet. Stories go the rounds of a near-naked Maasai in a cloak with spear and sword tending cattle in the bush who replies to some query in perfect English with an Oxford accent, and turns out to have a university degree. But the number of Westernised Maasai who have defected to the nomadic life is small.

What most threatens their traditions and their way of life is the ever-growing demand for land by people of the cultivating tribes. Much of Maasailand is unsuitable for cultivation, but agricultural techniques constantly improve. Already wheat is being grown on a considerable scale in the Kenyan part of Maasailand and, even more significantly, 'range management schemes' promoted by international agencies are introducing fences, rotations, dips and modern forms of livestock improvement which must spell the end of the nomadic life. 'Our spear-points are now like the teeth of infants,' Tepilit Ole Saitoti has sadly written. It is useless, he maintains, to weep over the death which he foresees for Maasai tradition and ways. His people, formerly conquerors, must learn to conquer new trends. Adapt or perish: that is the rule.

# Olduvai Gorge

Zebras, wildebeests and Thomson's gazelles graze on the plains near Olduvai Gorge where numerous remains of early man have been found together with the fossilised bones of the creatures that lived on the Serengeti two million years ago.

There is a good case to be made out for describing Olduvai as the second most celebrated gorge in the world, if we give pride of place to the Grand Canyon. Yet it is only about twenty-five miles long, and nearly all the fossils that have made it world famous have been found in the lower fourteen miles, where its steep sides reveal a series of strata covering a time-span of about two million years.

The Gorge could almost be said to start within a few hundred yards of Hugo's camp. When heavy rains fall, Lake Ndutu overflows into a small lake called Masak, and thence into the Gorge, which runs eastwards, getting deeper as it proceeds. Layer after layer of volcanic ash deposited here over the ages provide geologists with a kind of chart enabling them to

probe the course of the evolution of mankind's precursors far back into the Pleistocene age and beyond. (The Pleistocene lasted from roughly two and a half million to 10,000 years ago.) Louis and Mary Leakey did not discover Olduvai, but they did unearth there many of the fossils that have changed so dramatically the scientific estimate of the age of the human species and of the hominids that preceded us in the family tree.

Louis Leakey died in 1972, but Mary lives there still, directing a team who continue to make important discoveries. To reach her headquarters, you follow a track parallel to the Gorge on its northern side before dropping down through rocks and tuffs (volcanic ash deposited under water) and the wild sisal which give the Gorge its Maasai name. We left camp early, and before long almost ran over a vulture which was so absorbed in devouring the innards of a wildebeest that it neither saw nor heard us, and went on tearing at the guts of the carcass. We could have leant out and touched it with a hand.

Vultures, like diplomats, follow an order of precedence when partaking of a meal. Seven species dispose of carrion in this part of the world. The first to satisfy their hunger are the Ruppel's griffon and the white-backed vultures, much the commonest kinds. They have long, flexible necks which can reach down to the soft tissues inside the rib cage. Then come the lappet-faced with bills specialised for tearing, who devour muscle tissue and parts of the skin. They are followed by the white-headed, who often arrive first but move away to wait their turn while the griffons and white-backeds swoop in. The three remaining species, hooded, Egyptian and the rare lammergeier, are general purpose scavengers, and move in when they can.

Vultures of all kinds, however, come last on the list of predators who share the feast. Lions generally claim first place at the kill, although silver-backed jackals sometimes nip in, grab a portion and make off before the lions can react. When lions have eaten their fill in come hyenas, although these powerful determined animals have been known to gang up and drive a lioness, even two or three, from the banquet. Moreover it is now known, through the work of Dr Kruuk and others, that quite often it is the hyenas who make the kill and the Kings of Beasts who come in as scavengers. So the order of precedence is by no means always observed. Jackals await their turn, although sometimes they push in to get a share while hyenas are still feeding. Finally the vultures, waiting hungrily in the wings, smother the remains in a feathery blanket. Marabou storks stride about on the outskirts, stabbing with long bills at any scrap of meat their

This leopard climbed a tree to keep its prey out of reach of lions and hyenas.

quick eyes detect, even sometimes on one which a vulture has seized and is in the process of swallowing.

Thus a single carcass may support a dozen species of predator before the insects get busy to demolish what is left. Finally the bones will replenish the calcium in the soil whose vegetation will be consumed by animals as yet unborn. Leopards opt out of the protocol chain by carrying their prey into the branches of a tree where it is safe from non-tree-climbing jackals and hyenas.

Big, heavy birds, some of them flightless, are a speciality of these open plains. The ostrich is of course the largest; we passed many small flocks of perhaps half a dozen birds and once, in the distance, saw a 'crèche': that is, an assembly of all the chicks from several nests gathered together under the care of one of the mothers, leaving the rest free from parental duties. Ostriches are the shyest of the big birds and stride away at a spanking speed when a vehicle approaches – not, however, to bury their heads in

It used to be fairly easy to find leopards on the Serengeti: this mother's coat is probably now being worn by someone who supports the primitive fashion of wearing animal skins.

A male ostrich leaves his nest, apparently a bit early and overly eager for the still distant female to take her turn on the eggs.

sand. This myth is thought to have arisen because nesting ostriches have been seen with their long necks stretched out flat on the ground.

Less shy are Kori bustards, famous for their courtship display. We passed one so engaged on the track to Olduvai. He was puffing out his neck feathers and folding back his tail until it rested on his back with the tail-tip touching his neck, an extraordinary posture that has the effect of turning the bird white all over, whereas when it is not displaying, the colour is mainly buff and grey. So far as I could see, he was displaying to himself; there appeared to be no admiring female within sight. These large bustards may be five feet high and are territorial in habit; perhaps he was telling other cocks to keep away.

The big birds of the plains are mostly ground-nesting – they have no choice if they are flightless – and take the minimum of trouble over their nests. They simply deposit their eggs on the ground, as plovers do. We passed a Senegal plover incubating her two brown mottled eggs beside the road. She stood up and raised her wings to a horizontal position, displaying their white undersides. This gesture would seem to draw attention to her nest, but the sudden appearance of a small white object almost underfoot no doubt startles a large mammal, and causes a wildebeest, zebra or even an elephant to swerve and so avoid stepping on it. Plovers are plentiful: we saw from the track at least six species, including the black-winged, spur-winged, Caspian, and the ubiquitous blacksmith plover – as always looking trim and immaculate.

We paused en route to inspect a curious feature known as the Shifting Sands, or *barkhan*, a sort of mobile sand-dune consisting of gritty particles of soil blown by a strong, persistent wind into a crescent-shaped mound. The particles, hot to the touch, are blown up the windward surface and over the edge of a sandy precipice on the leeward side, and so the *barkhan* slowly advances. With the steady hail of sand come dung-beetles in their tens of thousands, whisked off their feet by the wind to be borne up the sandy slope and tumbled down the precipice on the other side.

Hugo had clambered to the top with his camera and I wondered why he was on his hands and knees. He was rescuing dung beetles in distress. Hugo has a respect bordering on affection for these beetles, for he considers them to be the saviours of the Serengeti plains. They are here in millions, and without them, he explained, large parts of this arid region with its light, friable soil would turn into a dust-bowl and get blown away. As it is, they roll the dung of millions of herbivores into neat round balls which they bury, and in which they lay their eggs. They also make another kind of dung-ball out of coarser material, which they eat. In this way the humus necessary for the vegetation is processed and preserved, and the plains survive. As a bonus, larvae which hatch out in the dung-balls provide food for jackals in dry seasons when larger prey have migrated away. So Hugo's consideration for these busy beetles has been fully earned, even if his rescue operation was fleeting.

It was extremely hot at the bottom of the Olduvai Gorge, the vegetation more than ever sparse and spiky. An earlier visitor, the geologist E. J. Wayland, wrote that 'every step is challenged by bayonets of sisal, and a hundred other needle-points impede one's progress. Almost every vegetative organ that can be modified to minimise water loss by transpiration has become a thorn'.

Olduvai Gorge has yielded a wealth of information on early man and the pre-history of the Serengeti. For most of the past two million years, the area looked much as it does today but once there was a large lake and fossils of crocodiles and hippos are common.

The stream that drains the Gorge had become a trickle and would soon be dry. When heavy rain falls, it swells to a torrent impossible to cross by car and hazardous to those attempting to do so on foot. Hugo once rescued a scientist who had, in his turn, gone to the rescue of an African swept away by the surging waters. Both were in trouble, and all that Hugo could see of the scientist as he went under was an arm rising from the water like the arm that seized Excalibur, but in this case clasping a pair of spectacles. Both men, and the spectacles, were saved.

We crossed the ravine and jolted on to Dr Mary Leakey's bungalow about three-quarters of the way up the opposite wall. All around were rocks, tuffs, thorn bushes and the wild sisal. Her simple dwelling, equipped with a pair of windmills to generate electricity, must seem the height of luxury compared to the hot and dusty camp the Leakeys shared for many years when funds were desperately short, but finds excitingly many. An underground tank to store rainwater caught by the roof must be the greatest blessing. In early camping days they had to use filthy water from a rhino wallow, unless they could afford the cost of the petrol needed to fetch water from a spring twelve miles away. An attempt to catch rain water from the fly of their tent resulted in a violent illness, for the canvas had been impregnated with an insecticide.

The Leakey family's association with Olduvai began in 1931 when Louis, then an unknown and penniless anthropologist, accompanied Dr Hans Reck to the Gorge. Dr Reck had led a German expedition to the spot in 1913 and had discovered fossil bones and a human skeleton. He had also identified four layers of volcanic deposits, labelled Beds I to IV, which range in age from about 17,000 years Before Present to some two million years B.P. Nowadays the age of fossils can be determined by radio-carbon dating if the fossil is no more than 60,000 years old, and by the potassium-argon process if it is older.

Lack of funds prevented the Leakeys from carrying out sustained research into these rich deposits for nearly twenty-five years. But they visited the Gorge whenever they could and uncovered a great many fossil bones and stone artefacts. It was clear that many early men and proto-men, hominids marking various stages in evolution between ape and man, had lived on these plains, and beside a vanished lake which then covered much of the region. But for years the discovery of a skull that would reveal what kind of hominids they were eluded the Leakeys.

The break-through came in 1959 when Mary, accompanied by two Dalmatian dogs, spotted an interesting-looking fragment of bone

Hippos exhibiting threatening behaviour during a territorial dispute.

protruding from the wall of Bed I, the oldest layer. It proved to be what the Leakeys had always hoped for and never ceased to believe would one day be found – part of the skull of a pre-human creature. Louis and Mary and their African assistants dug out some four hundred fragments which Mary pieced together with infinite patience. 'That took me eighteen months,' she said. They named the owner of the skull *Australopithecus boisei*, but the press preferred the nickname Nutcracker Man, and the Leakeys themselves affectionately called him Dear Boy. He was not a direct ancestor of *Homo sapiens*, but belonged to a side-branch of the human tree which died out perhaps about one and a half million years ago. This individual had lived, the Leakeys estimated, about one and three-quarter

The most exciting discovery made at Olduvai Gorge was the fossilised jaws and teeth of *Homo habilis* – one of man's ancestors.

million years BP. The name *boisei* was to commemorate a London businessman, Charles Boise, who had come to the Leakeys' rescue when they were in dire financial straits.

*Australopithecus boisei* brought fame to Olduvai and proved, nearly two million years after his death, to be a money-spinner. Digging accelerated, more and more fossils came to light and the bones and teeth of more hominids were found. The first major find, made by the Leakeys' eldest son Jonathan, was assigned to the species *Homo habilis*, so called because he was a handy maker of stone tools. He lived about two million years ago and Louis believed him to be in the direct line of modern man's ancestry.

Louis' belief that more than one species of hominid, and also *Australopithecus*, co-existed on earth was a revolutionary idea to many of his fellow palaeontologists. He was throughout his life an enthusiastic upsetter of apple-carts and challenger of sacred cows. Mary is an equally dedicated disciplinarian who checks and re-checks every supposition and jumps to no conclusions. It was a fine combination.

Since Louis' death, Mary has lived in the Gorge with her dogs, continuing the work and now and then going on lecture tours in North America. Deferred to as a world authority on the fossil record of mankind's ancestry, she prefers the relative peace and solitude (though she has many visitors) and hot sunshine of this familiar thorny ravine to the lecture platforms and social throngs of Europe and America.

Now the mantle of discovery, together with the gift of communication and a generous draught of his father's luck, has passed to Richard, whose own Aladdin's cave of riches lies in the sedimentary deposits bordering Lake Turkana in Kenya's arid northern province. Here, in 1969, during a reconnaissance on camels, Richard and Meave, a colleague later to become his second wife, came upon a hominid skull literally sitting on the surface of a dried-up river-bed – 'staring at us,' Richard wrote, 'a truly extraordinary moment.' The skull, they reckoned, had been embedded in the bank of the stream, the bank had gradually eroded away, and only a few months before their reconnaissance, and since the last heavy storm, the skull had rolled from the position it had occupied for about two million years to await, it seemed, in this bleak desert, the arrival of Richard and the future Mrs Leakey. The next heavy storm would almost certainly have washed it down into the lake's oblivion. It belonged to *Homo habilis*, the same species of hominid excavated at Olduvai about ten years before.

More was to come. Another skull of *Homo habilis* was found; this time by a member of a team led by Kamoya Kimeu, who had worked with the Leakeys since 1960 and become a skilled and sharp-eyed finder of fossils.

Unlike the former prize, and much more normally, it was dug out from the sand in hundreds of pieces. Painstakingly reassembled, it was labelled 1470 and delighted Louis on his last evening in Nairobi before he departed on a lecture tour from which he never returned. He died in London from a heart attack. This hominid was older than the Olduvai specimen, dating back more than two million years.

The Lake Turkana deposits continued to yield an ample harvest. In 1978 there came to light a skull belonging to a species rather later in the evolutionary scale: *Homo erectus*, who as his name implies walked upright, had a larger brain than *Homo habilis* – though not as large as our own – and a less ape-like face. He lived between a million and a million and a half years ago, and had been described by the Dutch scientist who had discovered the first specimen in Java in 1891, and named it *Pithecanthropus erectus*, the 'missing link' between man and ape. This is now dismissed as a simplistic idea; the ancestry of man is seen as a tree with several branches, rather than as a ladder up which *Homo sapiens* climbed from ape to man. The Leakey theory is that *Homo erectus* evolved in eastern Africa around a million or more years ago, spread into Asia and Europe, learnt to adapt from tropical to temperate conditions, developed skills in the making and use of stone tools, and began the process of subduing the environment which distinguishes the genus *Homo* from all previous hominids and from the apes. Then, somewhere about 300,000 years ago, *Homo sapiens* appeared. Just how and when and where is a matter for much discussion and dispute among archaeologists, and probably always will be, since the theories advanced to trace the origins of man rest, literally, on scraps of evidence. Scraps not only of the hominids themselves, but of the tools they used. Olduvai abounds in such stone artefacts, and sites near Lake Turkana have yielded evidence that tools were being made there some two million years ago, long before modern man appeared.

After coffee, Mary led us to her small museum to show us a cast of one of the most exciting archaeological finds ever made. It is a cast of a set of footprints impressed upon volcanic ash by three individuals, probably a man, a woman and a child, discovered at Laetoli, thirty miles south of Olduvai. The man walked in front, the woman followed mainly in his footsteps and the child to one side. Possibly the child held the hand of one of the parents. There are 31 footprints made by the adults and 39 by the child. The height of the adults, calculated from the length of their stride, has been estimated at 4 feet 7 inches in one case and 4 feet 1 inch in the other. The foot of a modern individual of about the same size can be fitted into the casts. The feet that made the prints were not prehensile like those of apes: they were flat with slightly splayed toes and were arched, like our own. And these beings walked upright.

These footprints were made, according to potassium-argon dating, around three and three-quarter million years ago. They were made in one brief moment – two or three minutes perhaps – on one particular day when, by an amazing chance, conditions were just right for their preservation, and lay there undisturbed until they were seen by their remote descendants nearly four million years later. Mary Leakey has described this find as the most exciting of her career.

The nearby volcano Sadiman must have been in eruption, spreading ash over the plain. This trio of hominids left their footprints in the ash as they walked along, and then it came on to rain – a light shower, not a downpour which would have washed the footprints away. Raindrops pitted the ash, and then the sun came out again and the ash dried quickly like cement, fixing the record – one is tempted to say forever, but nothing is forever and we must be content with nearly four million years.

It is even possible to hazard a guess at the state of mind of this little hominid family. Sadiman was rumbling and belching out ash, and probably other volcanoes were active too. Animals were all about and also left their tracks in the drying ash – not only hares, guinea-fowl and antelopes but rhinos, buffalos, elephants and a sabre-toothed tiger. The hominids were walking close together and may have been holding hands. Were they frightened? It would be surprising if they were not. No stone tools have been found that date as far back in pre-history as the era when this trio took their walk. So they were defenceless against predators, and against the unknown forces that dwelt in the volcanoes. Chimpanzees will hold on to each other when frightened – the sense of touch reassures. So it is, for that matter, with modern humans. So probably they were three frightened hominids walking over the plain.

Were they our ancestors, or did they belong to a hominid species that has died out? This is a matter for argument which could only be settled if a skull came to light and could be pieced together and linked to these free-striding, erect individuals. Mary Leakey's opinion is that they belonged to a primitive species of *Homo*, probably ancestral to *Homo habilis*, and if the Leakey theory is correct, remotely ancestral to ourselves.

The Leakeys had looked over Laetoli as long ago as 1935 and knew that fossils were there, but had no time or resources to undertake a dig. It was not until 1974 that a chain of events leading eventually to the footprints was started by Hugo's son Grub, then aged eight, and spending the summer holidays with his father. Not far from Hugo's camp at Ndutu is a pleasant tented lodge for tourists, which was at that time run by a stalwart

character called George Dove, renowned for his joviality and a fine moustache with pointed extremities.

He had spread some gravel taken from a nearby river-bed over his parking yard. One day Grub spotted a piece of fossil bone in the gravel. Hugo took his son to the river-bed and found more fossils, which they showed to Mary. Some time later, she and George Dove explored the dry river-bed which led to Laetoli and found a hominid tooth. Further investigations followed, rewarded by the discovery of fossil remains derived from twenty-six hominids, all in small pieces and including teeth, but no fragments of skull. A great array of fauna had also left bones in the river-bed, including a huge pre-elephant, a three-toed horse *Hipparion*, ancestral white rhinos and buffalos and three kinds of giraffe; three kinds also of hyena, a large sabre-toothed cat, and a host of smaller creatures such as we know today – dikdiks, mongooses, porcupines, rodents, monkeys, snails, even a clutch of guineafowl eggs. The hominids of Laetoli's Footprint Tuff, as it is called, must have had plenty to eat, assuming they could catch and kill the animals without weapons.

Dikdiks, pigmy antelopes the size of a hare, usually live in pairs in bush country.

# Ngorongoro Crater

Down at the bottom of the enormous Ngorongoro Crater lies a shallow lake encrusted with flamingos. Like all the flamingo lakes of the Rift Valley, it is alkaline and undrinkable. The flamingos need access to fresh water once a day, and make their way to the mouths of several small rivers that flow into the lake to drink and to wash the soda off their long red legs and beautiful pink and crimson feathers. The birds are so closely packed together you would not think that another could be squeezed in; but still they come, alighting in the shallows, ruffling their feathers, dipping their bodies in the water and evidently enjoying their bath.

Bolder birds venture a little way upstream to enjoy the pure water, but not far, and we could see why. I counted ten jackals lying on the low banks of the river, spaced out like sentinels, watching and waiting. Any over-bold bird would soon fall victim to a quick pounce. Hyenas also were keeping an eye on the situation. A couple came down to the water's edge and the birds retreated, chattering nervously. The hyenas walked up and down, as if on patrol. Then one of them waded into the water and the birds again retreated. At a further move from the hyena they took off in a great crimson cloud, circled round, and alighted again at a safe distance from the enemy.

There are other predators of which flamingos must beware. Fish-eagles swoop upon them; and marabou storks cause havoc by driving the frightened birds from their nests.

Flamingos build mud castles about eight to twelve inches high on which they incubate a single white egg; but quite often, after building great assemblies of these nests, they depart in a body without using them. They have a choice of soda lakes to feed in, but normally breed only in one, Lake Natron. They move about between the lakes in great flocks, following no regular pattern. Their movements appear to be governed by the water levels of lakes, which in turn determine the concentrations of the algae on which the birds feed. These are the lesser flamingo, 'East Africa's special bird' in the words of Leslie Brown. Bernhard and Michael Grzimek, father and son, surveyed the breeding grounds on Lake Natron from the air. They made a hole in the floor of their small aeroplane, fitted a camera into it and took a sequence of photographs which, when strung together, enabled them to count the birds. There were 163,679. Three months later there were fewer than 8000. These numbers, for flamingos, are small. I have seen over one million – or so I was assured – on Lake Nakuru, and have also seen that lake without a single one.

They do not breed on the Ngorongoro Crater lake, but we did see a

Often up to a million flamingos feed on algae in the Ngorongoro Crater.

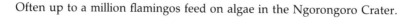

group displaying, part of their pre-mating ritual. They were greater flamingos, often found in small bands among a host of lessers. They ruffled out their feathers, spread their wings and marched shoulder to shoulder to and fro in the shallows, moving their heads from side to side as they paraded. Leslie Brown compared displaying flamingos to moving chrysanthemums. But the glory of the flamingo is in its flight.

It is often said that Ngorongoro is the largest crater in the world, but this is incorrect; it is number six on the list, the three largest being in Japan. It is however the most complete and spectacular, the others having partially collapsed sides or being mainly under water. Strictly speaking, it is not a crater but a caldera, resulting from the collapse of the volcano's cone.

To reach the Crater – as everyone calls it, correctly or not – you climb steeply through a belt of forest to an altitude of nearly 8000 feet, and here you breathe a different air. It is cool and fresh and pristine and has a tang of moist earth and humus and fallen leaves, with a hint of the aromatic fragrance of cedar-wood and wood smoke and the friendly smell of cattle. The Maasai pasture their flocks and herds in the forest glades. As the road corkscrews up the mountain you will see *manyattas*, or *engangs*, and come upon little humped cattle herded, as a rule, by small boys, and pass women, tall and upright, with their wide, bright bead collars and their nondescript brown robes.

There is anxiety among conservationists lest goats and cattle erode these slopes which are the source of rivers, and lest firewood cut for the *manyattas* will progressively denude the land of trees. From the conservationists' point of view, and in an ideal world, all livestock and human occupation would be excluded from this belt of forest in order to protect the streams. Then where would the Maasai go? It is not an ideal world, and the Maasai stay. They were here before the conservationists. But now their numbers are increasing, and with them the numbers of their cattle and goats. So pressure on the forest mounts.

When I was last on the Crater floor in 1963, there were *manyattas* there, and herds of Maasai sheep, goats and cattle. There was anxiety lest the flocks and herds should multiply to a point where they would over-graze the pastures and force out the wild animals. More recently, the Maasai have agreed to keep their livestock out of the Crater, so long as they can make full use of the highlands which encircle it. So it is here, on the Crater's rim, that the greatest danger lies of destruction of forests and the consequent drying up of springs.

The road to the Crater passes a simple stone memorial to Michael Grzimek, aged twenty-five when a vulture flew into a wing of his aeroplane and he crashed to his death. He and his father, Dr Bernhard Grzimek, then director of the Frankfurt zoo, initiated the system of aerial counting of the Serengeti animals which is essential for the management of the Park and Conservation Area. Their system, developed and refined, continues, and an annual count is made.

Finally you come to the view. About 2000 feet below, the Crater's floor lies spread out before you, stained by cloud-shadows, its lake shining like quicksilver in the sunlight. Concentrations of animals, isolated trees, rocks and boulders appear as tiny black specks on this vast parchment-coloured scroll. Away on the other side rise more mountains, dark with forest, steep and tumbled and remote. The floor itself covers just over one hundred square miles.

The German explorer Dr O. Baumann was the first European to come upon this breath-taking scene. He called the Crater 'an oblong bowl', rather a slipshod description for a German scientist, since a bowl is by definition round, but one can see what he meant. This was on 18 March 1892. He and his safari, escorted by numerous famine-stricken Maasai, whose cattle had been killed by rinderpest, clambered down the precipitous cliff, pitched their tents at the bottom, and lost no time in slaughtering some of the vast quantities of animals they found there. 'Although I am not at all a great Nimrod,' Dr Baumann wrote, 'during the day I shot one wildebeest and three rhinos; the latter we left to the Maasai.'

Rhinos were here in the Crater in such numbers that they were almost like cattle. Dr Baumann was sceptical of hunters' tales of the ferocity of these beasts, and of the hunters' courage in despatching them. 'Hunting them is not nearly so difficult or dangerous as it is claimed to be by professional Nimrods,' he wrote. 'The rhinoceros is not very shy, and if the wind is favourable one can easily approach to within thirty paces. To hit a rhinoceros at thirty paces you do not have to be a spectacular shot, and if the bullet hits the chest or (with a smaller calibre gun) the head, the animal usually collapses without further ado.' If only wounded, he continued, the rhino usually runs away, but if it should charge 'this moment is one described with vivid horror by the Nimrods. The companions flee and only the hunter bravely faces the charging colossus. This sounds terribly dangerous but the "charging colossus" is nearly blind and one step aside is sufficient to make it miss, and it charges past. When it stops and looks around for its enemy the hunter has plenty of time to kill it with another bullet at close range.'*

* These quotations are from *Ngorongoro's First Visitor*, a booklet of extracts from Dr Baumann's *Through Masailand to the Source of the Nile*, translated by Mrs G. E. Organ and annotated by H. A. Fosbrooke. East African Literature Bureau, 1963.

It was the same story everywhere – the hunters of the day slaughtered these clumsy animals in droves. A Captain Willoughby of the Indian Army, safari-ing with two friends on the eastern slopes of Mt Kilimanjaro, killed sixty-six rhinos. Count Teleki did better (or worse) with ninety-nine. Mr Henry Fosbrooke, commenting on Baumann's journal, suggests that profit as well as sport – if shooting rhinos can be so called – was a consideration. Even in those days rhino horn fetched good prices. Despite his protestations that he was no Nimrod, Baumann shot one of several 'magnificent snow-white' rhinos. These were not albinos but ordinary black rhinos who had wallowed in white alkaline mud.

How many rhinos are left in the Crater? A count made in 1980 put the number at less than thirty. A year later, I was told that a more recent count

It was thought that the Ngorongoro Crater would prove to be a relatively easy place in which to protect the rhinos but unfortunately, for various reasons, its numbers are still decreasing rapidly.

came out at twenty-two, and that, since then, three had been found shot (not speared) with their horns cut off. We saw two live rhinos lying together in the open and apparently sound asleep.

Despite relentless harassment, the rhino's unsuspicious vulnerability does not seem to have changed since Baumann's day. Even here, in open country and, in theory, under close protection, the rhinoceros seems unlikely to survive.

In 1963, nearly all the Crater's lions had succumbed to vicious attacks by a blood-sucking fly called *Stomoxys calcitrans*. Heavy and prolonged rain had caused the lake to flood over a great part of the Crater's floor. This had provided swampy breeding grounds for the flies which multiplied to such an extent that they killed their victims by stinging them and sucking their blood. (A single pair, an entomologist calculated, could give rise, in six months, to one hundred million million descendants.) The lions attempted to escape by climbing trees, but in vain. They tore their hides to ribbons by rubbing against trees and rocks to ease the irritation. An estimated lion population of sixty was reduced to fifteen. By 1980, if estimates were right, they had recovered only to the extent of doubling this number.

But it seemed that they were now making up for lost time. We counted nine cubs with two lionesses lying on and beside the road in various attitudes of relaxation. They did not budge an inch to make way for us. A little way back we had passed a lion and a lioness stretched out on a rock side by side and fast asleep. Hugo thought they might be courting; if so, it was a very placid courtship. An air of lethargy prevailed, as among the Lotus-eaters dwelling in 'a land where it seemed always afternoon'. However, the remnants of a zebra carcass suggested that the lions do eventually wake up and seek out a meal.

This pride was lying beside a winding strip of trees, bush and reeds fringing the path of the Munge river. A one-roomed cabin formerly stood under a fig tree near the pride's resting place. Here Jane, Hugo and Grub had lived, on and off, for two years while studying and photographing spotted hyenas and golden jackals. A pride of lions – very likely, Hugo thought, parents or grandparents of those lying on the road – had shared with the human party the shade and coolness of the trees and reeds around the cabin. On one occasion, it was found that Grub had been playing within ten paces of several lionesses lying in the reeds. On another, when Hugo was approaching the earth closet, there was a roaring and a grunting and the flimsy edifice was shattered as a lion who had been sheltering there took his leave.

'Those are the Scratching Rocks,' Hugo said, pointing to a low bouldery hill. They were so named because zebras sometimes rubbed itching hides against them, and the name was attached to one of the eight hyena clans inhabiting the Crater. Jane and Hugo's camp was in the territory of the Munge river clan, near its boundary with the Scratching Rocks clan. Many was the time that Jane observed adult hyenas, led by the dominant male or female, mark the border of their territory by rubbing a gland near the anus against a leaf-blade or twig as a warning to members of other clans – 'thus far and no farther'.

No system of passport control and network of Customs officers could be more effective in demarcating and observing territorial boundaries. When individual hyenas crossed over into the territory of another clan they were apt to show symptoms of uneasiness, and to retreat hastily if the rightful owners challenged them. Jane described in *Innocent Killers* a pitched battle between the Scratching Rocks and Lakeside clans, each side thirty or forty

Female wildebeests are no match for hyenas, so during the former's calving season the hyenas have an easy time.

strong, which surged to and fro along the boundary, first one side advancing and then the other amid a cacophony of whoops, growls and chuckles. The battle lasted for twenty minutes before the belligerents of both clans retreated to within their own territories, leaving one hyena dying of wounds and others severely mutilated.

Bloody Mary and Lady Astor were the two dominant females of the Scratching Rocks clan, which embraced about sixty members. Lady Astor was Bloody Mary's constant companion and friend. Both were portly, powerful hyenas. Bloody Mary had borne twins, Cocktail and Vodka, and Lady Astor gave birth to Miss Hyena, so called because she was seen staring at her own reflection in a pool of water, and because she was exceptionally well-favoured, with dainty ears, sparkling eyes and glossy chestnut spots. Wellington was the dominant male, with Nelson next in hierarchy. The sexes are difficult to tell apart, especially in youth; even Jane sometimes got it wrong. This characteristic gave rise to the false legend that they are hermaphrodites.

During the wildebeest rutting season the bulls become aggressive and will occasionally even threaten a car and commonly chase any hyenas in the vicinity.

Mrs Brown was a motherly type devoted to her single cub called Brindle, and she, like Bloody Mary, had a friend and companion, Baggage, possessed of a pendulous belly and liquid brown eyes. Her cubs were Sauce and Pickle. Quiz was a young male who fraternised with members of the Lakeside clan without incurring the wrath of his fellow clansmen – a very unusual situation. The separate personality of each hyena emerges from this study as clearly as do human personalities in a good novel. The image of the mean, cringing, cowardly creature is shattered for good. Not that all hyena habits are admirable. They can be savage in attacks on their own kind and their favourite meal is wild dogs' dung. But there is no doubt about their strong family loyalties, their ingenuity and a certain *insoucience* which sets them well apart from most other animals.

Hugo's golden jackals seem, to me, much more attractive animals. They are the fox of Africa and Asia, and when sport-loving Englishmen took packs of foxhounds to India, Kenya and elsewhere in colonial days, the jackal was the orthodox quarry, although the hounds, who must have been somewhat baffled by the variety of strange smells they encountered,

Left: Like fingerprints in humans, the stripes of each zebra differ.

Below: This young foal shows the long hair and brown stripes typical of infancy.

sometimes chased other animals such as small antelopes, wart-hogs, cheetahs and even, on one occasion, a leopard, which leapt up a tree. In the Crater Jason and his mate Jewel, with their cubs Rufus, Amba, Cinder and Nugget, occupied a hunting range of about one square mile. Unlike hyenas, but like wild dogs, jackals feed their young by regurgitating. Jason and Jewel hunted together and brought back ample food for their playful cubs, who stayed close to the den. Snakes, rats, spring hares, ripe figs and even mushrooms were included in their diet. After eating a particular kind of mushroom Rufus, the most adventurous of the cubs, appeared to go mad. He rushed round in circles and charged full-tilt at an adult wildebeest. Possibly he had hit upon a hallucinogenic mushroom. Unlike hyenas, jackals do not live in clans. The unit is the nuclear family and, almost certainly, they mate for life.

During one journey on the Serengeti we passed four jackal cubs playing near a den beside the road. They were chasing each other in circles, pouncing on imaginary prey, rolling each other over and then sitting still with pricked ears and enquiring eyes. A little way off their mother lay at rest. Hugo stopped the Land-Rover to examine her through binoculars. Some time ago, he said, he had watched and photographed a family of golden jackals in this same area, and she might be the same vixen. If he was correct, she would have a slit in her right ear. He moved the Land-Rover to a position from which her right ear came into view. Sure enough, it had a slit. The same vixen, in the same place, six years later.

The territories of jackals seem, as a general rule, to be remarkably constant. In six years, Slit-ear must have raised a lot of families. What had become of them all? Mortality among cubs is high. When Jason and Jewel produced a second litter, all five cubs died. Surviving males most probably go off to seek new territories, but the females, Hugo believes (unlike wild dogs), often stay near each other and keep in touch.

There is no proof of this, but behaviour has provided clues. Jackals guard their territories jealously and permit no strangers to approach. Yet Slit-ear, with her cubs in tow, was seen to trot over to the den of a neighbouring jackal, also a mother of cubs, and, as it were, pay a call. She was greeted not only without hostility but with friendliness. The father of the resident cubs lay down and went to sleep. The most likely explanation of this conduct was, Hugo believed, that Slit-ear was the mother of the jackal she visited. It seems likely that family ties, especially between females, continue to be recognised for longer, and are more durable, than has been generally realised.

Using air thermals, vultures effortlessly gain height and then glide long distances.

134

One day, near Seronera, Hugo and Jane watched an Egyptian vulture pick up a stone in its bill and throw it at an ostrich egg. After several attempts the shell cracked, and the vulture moved in to gobble up the contents. Was this, they wondered, an isolated, accidental event, or was it part of the bird's normal behaviour? No one, as far as they knew, had previously recorded it. They kept several Egyptian vultures under observation in the Ngorongoro Crater and soon saw the stone-throwing repeated. These vultures – and no other vulture species, it was subsequently found – were using stones as tools. This was an exciting discovery because tool-using had only been established once before in the case of a bird.

Jane and Hugo carried out experiments to find out what triggered off the vultures' stone-throwing behaviour. Different sized eggs, ranging from a hen's egg to a huge one made out of fibre-glass, and eggs painted in different colours, were put out for the vultures to tackle. Size and colour made no difference, but the shape did. So long as the object was egg-shaped, Egyptian vultures threw stones at it. The fibre-glass egg must have been very frustrating. The conclusion was that this behaviour was at least in part learned rather than inherited. Young birds observed their elders cracking eggs and copied them. It is odd that no other species of vulture seems (so far) to have adopted this procedure, and that it has been observed in the Egyptian species all over its range: in Kenya, Ethiopia, Sudan and elsewhere. Was it, in the first place, an accidental discovery that other Egyptian vultures had the wit to repeat on purpose? And why just Egyptian vultures? Another puzzle.

In a rose-red sky, the sun dipped behind the Crater's rim and the great bowl was drained of light and colour. Zebras no longer shone, their stripes blended into grey shapes moving on an ashen plain. The hills darkened to the colour of a ripe grape and then to charcoal. We came to a grove of fever-trees (a species of acacia with reddish yellow bark that favours damp conditions) inhabited by vervet monkeys, and heard hippos grunting in a swampy lake under the trees, and saw a waterbuck standing nobly with his head raised and the white ring on his muzzle showing in the twilight.

Then came the climb out of the Crater. From below, you can see no way by which a road might spiral up this sheer forest-clad cliff. Somehow a road does make the ascent, though it is rough and narrow with one-way traffic. Hairpin bends often have to be negotiated in two stages.

We came to forest and to darkness, a canopy arching over our heads and a skimble-skamble of disorderly, moss-hung trunks and branches proliferating all around. Such dark forests arouse in all of us, I think, vestigial fears of malignant monsters and werewolves and witches and

Egyptian vultures using 'tools' to break open an ostrich egg.

Overleaf: This peaceful scene belies the fact that hippos are, at times aggressive. It is almost certain that more people have been killed by hippos than any other large creature in Africa.

things that go bump in the night. They also aroused in me some apprehension lest we should encounter the elephants who live here, or for that matter any other animal who would cause Hugo to switch off his headlights. Less than a foot separated our wheels from the edge of a precipice and a sheer drop of about a thousand feet. It was not so much the elephants, as Hugo's habitual courtesy towards all animals that was the main cause for concern.

As the Land-Rover clambered up, night having fallen, we heard crashes ahead and, rounding a bend, saw dark shapes above us on the bank on our left-hand side. More crashes, and dark shapes were on the road ahead. Sure enough the lights went out but, thankfully, the brakes went on too,

and we stayed put. Not elephants, but a herd of buffalos going down the mountain side and crossing the road. One could sense rather than see them, deeper concentrations of blackness, for the moon had not yet risen and the forest canopy came between us and the stars.

Buffalos, generally speaking, have placid natures but, like all herd animals that move about in phalanxes, they are inclined to expect other animals to get out of their way. I hoped they had good night-time eyesight and would not mistake our Land-Rover's roof for a tree-trunk or boulder. For such large animals they are, in fact, very nimble, and they jumped down from the bank and crossed the road without putting a foot wrong. Their strong bovine smell hung in the air as we drove on and up to the Crater's rim, and to the welcoming lights of the comfortable lodge. Nights are chilly, and justify a fire of aromatic cedar logs in an open fireplace.

Presently a half moon was reflected faintly from the lake below. The plain around was silvery too, with dark lines marking river-beds, and shadows where the folds of the hills flattened out on to the Crater floor. Down there the night was daytime for nocturnal creatures and for predators who would be about their business. For dinner we ate meat that someone else had killed; they killed their own. I hoped that the rhinos, defenceless and short-sighted, would live through the night.

Next morning I watched the sun rise over the mountain ranges on the Crater's far perimeter, a great tumble of peaks and old volcanoes rising to nearly 12,000 feet. Lemon yellow deepened to orange and to a molten red streaked with purple. Then the sun in splendour burst from behind the mountains and hurled long spears of light across the Crater. The lake below shone like a new shilling, specks that were trees and rocks and grazing animals sprang to life and something winked far below, moving across the plain. Binoculars showed this to be the windscreen of a vehicle. There is a camp-site in the Crater, and I knew who owned the winking vehicle – Steve and Sharon, to whom we had said goodbye at Ndutu. A week's respite from flying had awaited Steve in Nairobi and, with Sharon, he had hurried down to Ngorongoro for a few more days among the animals. We had run across them the day before among the flamingos and asked them what they'd seen. 'Everything', was the happy answer.

A lawn as green as an Irish meadow surrounds the lodge's cabin and the central dining room, and grazing on it were two large and very fat buffalos, both bulls. No wonder they were fat; compared to the dry, sparse grasses of the Crater, this rich green Kikuyu grass was like a fat duckling to a withered sparrow. Statistics show that many more buffalos die from malnutrition and starvation than from any other cause. Not these buffalos. They had been clever as well as bold, and I wondered why all their aunts and uncles, brothers and sisters and cousins, hadn't joined them. Perhaps these were the pacemakers and in time the rest would follow, which might prove an embarrassment to the human occupants of the lodge.

One of the buffalos was rubbing his neck against the back of Hugo's Land-Rover, which was parked outside a row of cabins. I went back to get my camera and passed within a foot or two, smelling his sweet grassy breath. He paid no attention. His eyes were mild and moist, his horns with their thick bosses made him look top-heavy, his hide was glossy. A tick-bird was perched on his withers. Both buffalos went on munching fresh young grass, the picture of contentment. Had they been dairy bulls in an English meadow, I should not have walked past them unperturbed and, if I had, they would not have ignored me.

Having a higher rainfall than the surrounding country, Ngorongoro Crater's 100 square miles provide lush grazing grounds for herbivores.

How attitudes have changed! In my youth buffalos were looked on as rifle-fodder. They were considered highly dangerous and classed as vermin – lions were in that category then. And, of course, for farmers this was true. Lions took your cattle, buffalos and nearly all the other ungulates trampled your crops, elephants broke down fences. Nowhere can farming and large wild animals happily co-exist. At one time, the government of Kenya issued ammunition free to farmers to help them protect their crops. Tens of thousands of zebras, wildebeests and other ungulates were destroyed.

When it came to sport, much argument centred on which of the 'big five' was the most dangerous. The 'big five' embraced the species that

Lone buffalo and herds containing two to three hundred are a common sight in Lake Manyara National Park, where they are the major prey for the resident lions.

sportsmen most wanted to shoot: elephant, rhino, buffalo, lion and leopard. All were believed to be *ipso facto* dangerous and liable to charge on sight. Each species had its champions when assessing its ferocity: for, naturally, the more ferocious the animal, the greater the bravery of the hunter who (armed with the last word in high-velocity precision rifles) pursued it at risk to life and limb.

The buffalo was often placed at the head of the list. This was because buffalos were said to be cunning and treacherous – though how they could have owed loyalty to people who were trying to kill them I could never, even then, understand. If a buffalo was wounded, it would often double back on its tracks, conceal itself in undergrowth beside the trail and charge its pursuer as, with eyes downcast, he followed its blood-spoor. I knew a young man who was killed by a buffalo in just such a way.

I don't think it ever occurred to anyone to question the belief that these animals of the 'big five' were by inherent nature savage and the enemy of man. The proverb *le tigre c'est un méchant animal, si l'on attaque, il se défend* was never quoted. The fact went unrecognised that, in the great majority of cases where an animal attacked a man, the animal had either been first attacked, or at least frightened, by the man. Very often, a so-called 'vicious' animal would be found to have suppurating wounds from an earlier encounter with sporting man.

There are, of course, exceptions. A female will defend her young against a threat, or what she believes to be a threat, in most circumstances; even a domestic goose will charge, wings flapping, if you approach her nest. Rogue animals appear now and then, such as man-eating lions or tetchy elephants. Animals in pain from some natural cause – one can imagine the agony of an elephant with tusk-ache – may vent their misery on anything that crosses their path. No sensible person takes risks with creatures so much stronger and swifter than himself. But the myth that these larger mammals are by nature hostile to man has, by and large, been exploded. Normally they are dangerous only in so far as they are provoked. Otherwise, when they see man or smell him, they simply run away.

At any rate, those two fat buffalos displayed no treachery or cunning, just placidity. The lodge's African staff started to trickle in to prepare the dining room for breakfast and to sweep the paths. They took no notice of the bulls, who went on munching. I offered up to them, the friendly bulls, a silent confession and prayer for forgiveness. I once shot a buffalo – two in fact. I was only sixteen, and buffalos were all around in those days and a menace to crops. But now I am sorry that I shot those buffalos, and, alas, other animals, before the weathervane of opinion swung from 'slaughter' to 'conserve and respect'.

# Lake Manyara

A pair of grey-headed kingfishers was nesting in the sandy bank of a stream that, bubbling up from a spring at the foot of an escarpment, flowed into Lake Manyara. Hugo wanted to photograph one of these lovely little birds at the mouth of the nest, just as it was about to fold its wings and dart into the hole. We sat in the Land-Rover about fifteen yards away, his 560-mm lens trained on the spot, and waited.

It was very pleasant in the bed of the stream, which had dwindled here to a chain of shallow pools and little swamps overhung by spiky branches. Vervet monkeys were foraging under the acacias and clambering in and out of a tall fig tree. Swallows swooped in graceful arcs to devour flies hatching in myriads in patches of swamp. A movement in the bush that fringed the water-course resolved itself into an impala doe picking her way down the bank, big ears twitching, tail a-flicker, followed by half a dozen companions. They sipped from a reedy pool and filed up the opposite bank, followed by the male with a fine pair of horns swept back

In a territorial display a grey-headed kingfisher flashes its colourful wings warning others, except its mate, to stay away.

over his shoulders. Impala males lead a harassed life, perpetually rounding up their females and trying to keep them from straying, and frustrating the designs of other males to oust them from their suzerainty.

One of the kingfishers perched on a twig on the edge of the stream, a scrap of brilliance in the sober-tinted bush, its azure wings and white breast shading into russet, with red legs and orange bill. It mate was flying restlessly about, never far from the entrance to the nest, but never actually approaching it. If we stayed there long enough, she would get used to us and doubtless think it safe to fly into the hole. A black-and-white fish-eagle flew in to perch on the topmost branch of the fig tree, which shook from time to time as the vervets leapt about. Several small furry creatures darted across and up the far bank – pigmy mongooses. There was plenty to see.

A pair of D'Arnaud's barbets was also nesting in the bank, a few feet away from the kingfishers, and they were more confident. In and out they went, flashing their red rumps, and Hugo shifted his lens to focus on their front door. No sooner had he done so than the kingfisher swooped down, hovered for a moment at the nest's entrance with wings fluttering photogenically, and shot down the hole. Hugo muttered softly to himself. Impatience had not been rewarded. He shifted his lens back but the kingfisher had evidently decided to spend the rest of the afternoon down the hole.

There comes a moment, indefinable and inexact, when afternoon, with its implication of siesta, the dead flat time of day, slowly turns towards the first of evening with its wakefulness and lengthening shadows and thoughts of water, as if the heartbeat of the day were quickening. Elephants begin to pace towards their favourite pools suitable for mud-bathing, buffalos drift out to graze, pelicans take off to seek their fish dinners, wheeling and circling above the lake. The fish-life in these waters must be rich beyond belief to sustain these tens of thousands of big hungry birds, and to support as well the human fishers who dwell around the southern portion of the lake, which lies outside the boundaries of the National Park.

Manyara spells elephants. It is not that there are more of them than in other parks – there are very many more in the Serengeti – but here they are much more concentrated and more easily seen. It is a small park, embracing only 129 square miles of which 89 square miles consist of water, leaving only 40 square miles of land – about the same size as Nairobi's park. A narrow strip of land less than thirty miles long and one to three miles wide stretches from one end to the other. On the west it is confined by an almost sheer escarpment, which is the eastern wall of the Gregory

145

Rift Valley, and on the east by Lake Manyara. A road runs from the northern gate through acacia woodland, between cliff and foreshore, to hot springs at the southern end of the park. Beyond that lie human settlements. You can drive from end to end and back in a few hours. But it would be a mistake to hurry.

This little park holds an astonishing variety of both animals and vegetation. Half a dozen or more different habitats are compressed into the area: the rough and rocky terrain of the Rift wall; a forest of tall trees and thick undergrowth fed by a high water table; acacia woodlands which have an orchard look about them; small areas of open grassland; streams trickling down from the escarpment; reed-beds; the lake shore teeming

Pelicans, in their thousands, fish on Lake Manyara and nest in trees in the National Park.

146

with bird-life and the lake itself teeming with fish. So, as a result, what scientists call the biomass – the total weight of live animals in a given area – is exceptionally high.

This is a place for loitering in with eyes peeled. There is always something stirring. A family of wart-hogs stands on the alert; father and mother and five piglets, their tails upright as telegraph poles, with tufts on top. Father wheels, and they trot away importantly in single file as if late for a meeting. A monitor lizard wearing the supercilious expression of all lizards lies sunning himself on the bank of a rivulet. A pied kingfisher (there are eight kingfisher species in this park) hovers and then dives so cleanly as scarcely to leave a ripple on the water, but misses his prey.

We halt beside a termite mound about six feet high and riddled with holes. At first I see nothing. Scrutiny through the binoculars reveals a small, pointed, bright-eyed face framed in the mouth of one of the holes. It is motionless and then disappears. A few minutes later, a ferret-like

Like other pigs, wart-hogs love lying in mud or water. Normally vegetarian, they occasionally eat carrion.

creature emerges cautiously from another of the many holes, runs across part of the mound and vanishes down yet another aperture. Soon others do the same; the mound is like a high-rise block of flats crammed with restless tenants. These are pigmy mongooses who, when on the move, scurry in small packs and, when they wish to take their bearings, stand upright to peer over the grasses, forepaws hanging down like limp gloves. To get the photographs he wanted, Hugo had once spent all day and every day for a fortnight following a single pack.

But we are looking for elephants, and soon come upon a family group standing under some acacias, relaxed and calm, gently swinging their trunks and now and again raising and lowering their great ears to cool themselves. The tusks of many of these Manyara elephants curve inwards until the tips meet and sometimes cross over. This must reduce their usefulness for stripping bark from trees and other purposes. Some elephants are completely tuskless. Could this be a response to centuries of ivory-hunting by humans? One might suppose that, by the process of natural selection, a tuskless strain would survive and eventually take over. It is not disputed that the average weight of ivory throughout East Africa has greatly declined. The big tuskers that so excited the admiration, and satisfied the greed, of former Nimrods (as Dr Baumann would have called them) have become virtually extinct.

These Manyara elephants will often allow vehicles to approach to within a few yards. Hugo had already spent several months photographing them and evidently they knew his Land-Rover by sight, and remembered it. They were not so happy about unfamiliar vehicles. On one occasion, when a group had been calmly advancing and browsing as they went, we saw them turn and hurry back. The sight of an unfamiliar 'Round Africa' bus, larger than the usual Volkswagens and differently shaped, had upset them.

Dr Iain Douglas-Hamilton spent the best part of four and a half years following the Park's elephants both in a Land-Rover and on foot. Since he and his wife Oria departed in 1970 others have intermittently continued his work. We encountered the current 'elephant man', John Sherlis, a young American from Harvard with the physique of a footballer, who was embarking on an eighteen months' field study.

The tolerance of elephants towards this constant invasion of their privacy is extraordinary. But, now and again, their patience wears thin. All they will usually do in this case is to stage a mock charge, or threat display. They raise their ears, wave their trunks from side to side and

Dwarf or pigmy mongooses are safe in the air vents of termite mounds.

gallop forwards, trumpeting fiercely. If you hold your ground, as a rule the elephant will halt when a few yards away and call off his threat, or vent his annoyance on some inoffensive object like a log or bush. But, now and again, he means business and doesn't halt. This happened once to Hugo, who only just escaped by driving all-out in bottom gear – the elephant was too close to risk a momentary check brought about by changing gear. The tusks of this elephant, a cow, had grown inwards until they were touching, point to point.

'There she is,' said Hugo, as we approached a family party under the trees. By the shape of her tusks, and the nicks and tears in her ears' edges, he had identified her, he believed, from Iain Douglas-Hamilton's photographs and drawings, as a matriarch called Sarah – one of a family of ten or twelve. Iain described Sarah as normally gentle and inoffensive, but on one occasion she charged his Land-Rover and plunged her tusks into its radiator. She had good reason; Iain had darted a young bull belonging to her family unit, probably a grandson, and had just leant out of his vehicle to administer the antidote. She did not press home her attack, nor had she done so when she put to flight Hugo and his vehicle. Now she showed no sign of ill-temper. Elephants often display uneasiness by rocking a foreleg to and fro, sometimes kicking up dust as they do so.

Hugo examined her more closely. 'The tip of one tusk has broken off since I last saw her,' he said. When two ingrowing tips meet, and growth continues, pressure must be exerted on the base of the tusks, on the elephant's equivalent of the gum. The nagging pain of a permanent gum-ache would aggravate the sunniest-tempered elephant, and I wondered whether this was the reason for her former irritability. Now the tips no longer met, and tranquillity had been restored.

For some time we watched a sparring match between two adults, one of them a young male. They advanced upon each other with upraised trunks and clashed their tusks together; one retreated, then returned to the fray. Tusks clashed again amid half-squealing, half-trumpeting challenges. This might have been one of the mock battles that quite often occur, or it might have been part of the process of expelling a young bull from his family group. When bulls attain sexual maturity, at twelve or thirteen years old, they are driven, or withdraw, from the family to lead a bachelor life sometimes in company with other males or, on occasion, and for short periods, with non-related female groups.

Elephants are gregarious creatures and don't normally seek out solitude. One of their most endearing habits is to touch each other with their trunks, often inserting the tip briefly in a companion's mouth – a sort of kiss. (We saw a youngster sucking the tip of his own trunk to reassure

Physical touch in greeting is an important element of the elephant's social behaviour and relationship with others. One gesture, putting the tips of their trunks onto each other's lips, is the equivalent of kissing.

himself, much as a human infant will suck his own thumb.) There are other similarities between the behaviour of elephants and humans – and differences too. Elephants, by and large, appear to be gentler, more tolerant, more orderly and, unless frightened, much less aggressive than man. The Chagga tribe, Sirili informed us, believe that long ago God sent a flood to sweep away evil-doers dwelling on Mt Kilimanjaro; down on the plains, the outcasts turned into elephants. Presumably they reformed, because the Chagga, in the past at any rate, would never hunt or hurt an elephant. 'You have only to look at elephants to know that they are human,' Sirili said. 'What other animal has a woman's breasts between its front legs?' The Kikuyu tribe, living over two hundred miles away around Mt Kenya, used to believe in this same legend. The sin for which the humans were punished, according to their version, had been extravagance and vanity – they had washed themselves in milk. An angry god transformed them into elephants who bore milk-white tusks to keep alive forever the memory of their folly.

Two family units of elephants, consisting of females with their offspring, each led by a matriarch, approach each other in Lake Manyara National Park.

Elephant calves need to learn how to use their trunks, here as a snorkel. While adults use their trunks to suck up water, than transferring it to their mouths, young calves bend down to drink with their mouths.

Elephants live in family units of some ten to fifteen related adults with their offspring, each led by a matriarch, generally the oldest of the group. A mature female may have three children still keeping her company: an adolescent, a younger child and a baby. These Manyara elephants follow a fairly regular routine. In the evening they move away from the flattish strip of land bordering the lake into the foothills of the Gregory Rift, which rise to about 6000 feet. Here they find shelter in rocky gullies, and an evening meal among the trees. Next morning, once the sun has risen over the hills, they make their way down again, feeding as they go, to their watering places in the streams that issue from the wall of the Rift.

While the waters of this lake are less alkaline than those of Natron and Magadi, they are still very bitter, and the freshwater streams are much

preferred by all the animals. After a leisurely drink, the elephants may take a mud-bath – which they obviously relish – and then resume their meal. (Except when sleeping, elephants seldom stop feeding.) When taking their siesta, the elephants look wonderfully relaxed in dappled shade in the heavy heat of noon and early afternoon, trunks hanging down, ears gently flapping to help cool the great bodies. After a couple of hours or so they move off for another drink and then a final feed near the lake before they make their way back once more to the lower slopes of the escarpment.

I know of no more pleasant way to pass the hours than watching Lake Manyara's elephants, who have grown so used to humans – or at least to

Dust baths, often taken after bathing in water, probably help to keep an elephant cool, as do its large ears. Blood flowing into the ears is cooled by 16°F before recirculating.

Land-Rovers – that they pay no apparent attention and accept you as they would any other animal posing no threat. But there is a fragility about this relationship. Any sudden movement or sound would fracture it, and the elephants would either be off in a flash or might decide to charge. Like all wild animals – or nearly all, I'm not so sure about lions – they are nervous and constantly alert, even when they seem most somnolent. Hugo recalls one occasion when the peace of a group at rest under a tree was suddenly shattered: the sound of trumpeting filled the air, the elephants took off and, bunching together, fled into the bush as if pursued by devils. He scanned the scene through binoculars but could see no sign of any strange or menacing object – just a sounder (troop) of wart-hogs trotting through the bush. On another occasion he saw the leader of a party that was ambling along suddenly halt, wave her trunk about, and change both speed and direction to lead her companions off at a tangent. A blacksmith plover had risen from her nest in the path of the group, raised her wings in

An elephant calf investigates by smell and possibly touch what its mother is eating, in this case soil which probably contains salt.

155

the distraction behaviour common to most plovers, and thus successfully diverted the herd away from her nest. Subsequent observations of elephant behaviour confirmed the belief that the sight of any running creature sets off alarm bells – even when the runners are only engaged in play.

We followed Sarah's family party on several occasions, watching them feed and rest and drink and have their mud baths. This was a fascinating sight. The adults sucked up mud and sprayed it over their backs and heads, whilst the youngsters plunged right in to the black ooze until all you could see of them was the tip of a trunk raised above the surface like a snorkel. Hugo was worried lest one of the adults should fail to see a submerged calf, and tread on it. But they are very careful. The mothers know exactly what their children are up to and will come to their rescue if one trips and falls, uttering as a rule an urgent squeal, or gets into any other kind of trouble. After the mud-bath, the elephants rubbed themselves against tree-trunks, or in one case against a termite mound, as if to alleviate itching. The whole process is perhaps a kind of skin massage. When they emerged from the mud pool they were black, but when they took a dust-bath they were brick-red.

An aspect of behaviour shared by elephants and humans is the habit of orderly queuing. Elephants are more polite than humans, or at any rate some humans, and queue-jumping seems to be unknown. I remember watching several hundred elephants coming down at eventide to drink from the Mudanda dam in the Tsavo East National Park in Kenya. You can watch this ceremony from the smooth grey saddle of a rock high above the dam, while the slanting sun throws long tree-shadows over the bush-speckled plain and reddens the dust-filmed backs of the great beasts. I thought that it was rather as if one of those diagrams of bygone battles, with each contingent of troops neatly arranged and classified, had come to life, with elephant groups replacing armies.

From bush and plain the elephants converged upon the dam in their family parties, dignified and patient, each group awaiting its turn. There was no unseemly scrambling, pushing or shoving. The members of one group advanced with even paces to the water's edge, drank, sprayed water over themselves and each other while the calves splashed and played just like human children. When everyone had finished, the group withdrew with leisurely precision and another took its place until all had drunk, darkness was falling, and the last party, still slow-marching with their calves beside them, melted into the bush. Without traffic lights or

When travelling from one area to another, elephants often walk in single file.

traffic wardens, without policemen, timetables or commanders-in-chief, they achieved the desired results far more effectively than those usually charged with directing crowds of people or bodies of troops.

Students of animal behaviour explain that the secret of this orderliness lies in the system of dominance which governs the relationships between family groups as well as between individuals within a family. Subordinate family groups will await their turn at the watering-place until all the members of a dominant family have slaked their thirst and completed their ablutions.

Though their numbers were smaller, the Manyara elephants behaved in just the same way as those at the Mudanda dam. We watched Sarah's party drink their fill before another family, who had meanwhile waited twenty or thirty yards away, moved in. Young of all ages, from a few months old to teenagers, ambled after their mothers while the calves played about underfoot, often getting in the adults' way. One in particular

Elephant calves love chasing birds in play. However, not all small birds fly off. Some, used to elephant calves, ignore the youngster who, losing its nerve, rushes back to its mother and, protected underneath her finds reassurance by sucking the tip of its trunk.

was making a nuisance of itself. He, or she – I think it was a she – kept on approaching other youngsters with invitations to play, waving her small trunk about and lying down with her legs in the air. Adults are extremely tolerant towards the young but when this party reached the pool they took a firm line and drank first, gently pushing away any calf that tried to precede them into the water.

There is something irresistibly comical about tiny elephants. When newly born, it seems, they do not always know what their trunk is for, or how to use it. Hugo watched one, not long born, who, running after his mother, inadvertently trod on the tip of his trunk. Instead of lifting his foot to free the trunk, he started pulling, rather like a blackbird tugging at an earthbound worm. The trunk stretched, the calf tugged until he was leaning over backwards, and when, finally, trying to keep his balance, he moved his foot away, the trunk, suddenly freed, shot upwards while the calf fell over backwards. It did not take him long to master the control of this versatile protuberance, and soon he was dashing about and waving it at birds: only at small birds, however. A bird larger than a wagtail would cause the calf to seek refuge underneath his mother's tummy.

'To me the death of an elephant is one of the saddest sights in the world,' wrote Iain Douglas-Hamilton in his enthralling book, *Among the Elephants*, written jointly with his wife Oria. He had no doubt that elephants, with their close family ties, feel the emotion of grief. He told of an occasion when, hearing the cry of a distressed calf, he clambered part-way up the escarpment to find a female lying with her feet wedged under a boulder and her head hanging down over a steep rock-fall. She had evidently put her foot into a pig hole, lost her balance and tumbled some four hundred feet to her death. Three of her children were standing round her still-warm body. 'The eldest', Iain wrote 'was moaning quietly but every so often gave vent to a passionate bawl. The second just stood dumbly motionless, its head resting against its mother's body. The smallest calf, less than a year old, made forlorn attempts to suck from her breasts.' The eldest calf then tried to move her body with his small tusks. If this is not a scene of grief and mourning, then what is?

It is well known that elephants will stay near a sick companion and try to succour it, sometimes by stuffing leaves into its mouth, and that they will try to lift it with their tusks. The most touching example of elephant devotion I have seen, though alas not witnessed, was a sequence in one of Alan Root's superb films, which depicted the tragedies of drought in Africa. Everything has withered, the river-beds are dry, but in some of them a little water can be sucked up by the elephants who thrust their trunks deep into the sand. An emaciated family party is trudging over the parched savanna in search of such a river-bed, the calves tottering by their mothers' sides. A calf collapses, and is too weak to rise. With their tusks the mother and another female lift it to its feet but it cannot stand. Plainly, the calf has died. But the others will not admit defeat and go on trying to keep it upright. These attempts continue, the commentary tells us, for four hours. Then, slowly, one by one the elephants turn and plod away to resume the search for water on the success of which their lives depend. Only the mother remains, undecided, beside her calf. She sees her family retreat, starts to follow, then looks back. She is faced with an agonising choice. 'There can be only one answer', says the commentator, 'for an elephant.' She remains by her dead child.

Elephants display an extraordinary interest, which as yet no one has been able to explain, in the bones of their dead conspecifics. Several instances have been recorded where a party has approached a carcass with every symptom of excitement, sniffed it all over, lifted the bones and twirled them in their trunks, carried them about and deposited them some distance away. They exhibit a special interest in the tusks, which they sometimes try to pull out of a carcass.

Elephant cemeteries are a myth, but not elephant burials. Other members of the group will sometimes pile branches on the corpse of a companion. Hugo has suggested that they are trying to conceal the carcass from the eyes of vultures who, if they spot it, may attract to the scene the elephants' most deadly enemy, man. But this is supposition, and no one really knows what lies behind this behaviour. There was even a case, recorded by George Adamson, of a Turkana woman who was buried alive. She was half blind and, in the fading evening light, lost her way and lay down under a bush to sleep. She awoke to feel elephant trunks sniffing at her body and lay there paralysed by fright while the elephants broke off branches and piled them on top of her. So thoroughly had they buried her that she couldn't free herself and was saved only when her cries next morning attracted a passing herdsman.

These elephants are relatively, though not entirely, safe within the Park, but far from safe outside it. Human settlements are spreading to the south and south-west, and on top of the escarpment they press against the forest's edge. Any elephant that sets foot outside may be, and many are being, legally destroyed. To the south lies an area formerly leased to European farmers who, attempting to protect their crops, slew the animals in large numbers. One can scarcely blame the farmers. One, an Italian, reckoned that he lost half his crops from elephant damage. He shot and shot and destroyed, before he tired of it, some five hundred elephants. Survivors who fled back into the Park naturally regarded humans as their enemies. Those in the southern portion of the Park, where bush and forest are much thicker, are a great deal less trusting and therefore more dangerous to man than those in the north. Even Iain and Oria Douglas-Hamilton, who almost became elephants themselves, learnt to dread the 'terrible Torone sisters', as Iain called them. There were four, and they charged on sight and charged to kill. One of the sisters was shot by a Park ranger after she had embedded her tusks in Iain's Land-Rover and was about to demolish its occupants.

What has happened to the other three? I am glad to say we did not meet them. Probably they have been shot. But I wished that we could have encountered Virgo, one of Boadicea's family, who became so friendly that she took fruit from Oria's hand. And when the human pair, their studies over, returned to the Park for a visit after an absence of two years, Virgo remembered them and let them walk up to her and touch her, and they, for their part, let her sniff at their baby with her trunk to satisfy her always lively curiosity.

Later, the Douglas-Hamiltons came to believe – as Jane had done in the parallel case of the chimpanzees – that it was a mistake to feed the elephants and beguile them into losing their fear of man. Manyara teems with tourists; they are not supposed to get out of their vehicles but sometimes they do, and it is easy to envisage a situation in which an elephant approaches a tourist hoping for a banana, is disappointed, and charges the man. Or the animal itself might get too bold, as happened in the case of an elephant in the Kabalega National Park in Uganda. Tourists fed him bananas from the windows of their cars and the elephant, naturally equating cars with bananas, took to lifting up the vehicles and shaking them in order to get at the fruit. The elephant had to be shot.

The tolerance of elephants, both towards each other and towards their ancient enemy, has been remarked upon by all who have got to know them as individuals and studied their social life. The territorial disputes that continually arise among many other species are noticeably absent among elephants. While each family group has its own range, groups will coalesce and mingle without acrimony, while each unit retains its cohesion and remains centred on its matriarch. Live and let live appears to be their basic philosophy.

As we drove along the lake shore, every tree was packed with nests of pelicans and cormorants. But for how long will the trees be there for them to nest in? All around lay evidence of elephant destruction. The worst sufferers were, as everywhere, umbrella acacias, *Acacia tortilis*. We saw acacias dead and dying all over the place. Iain Douglas-Hamilton had

Elephants spend most of the day and night feeding.

selected and marked some three hundred *tortilis* trees for study. When he returned to the park a few years later, over one-third of them had been killed by the stripping of their bark. There can be only one end to this. Like the humans around them, natural increase is swelling the numbers of the Manyara elephants and, added to that, elephants are coming in from areas where they are poached or hunted. Elephants can no longer trek away from an area where vegetation has been denuded to other regions where food is abundant, thus allowing natural regeneration of trees and bush to take place. There is nowhere for the elephants to go. They are bottled up, and threaten to destroy their own habitat. Sooner or later many of them must either break out of their bottle and be shot, or die of starvation.

Everyone concerned – scientists, park staff, tourist operators – is only too well aware of the problem. Various solutions, or attempted solutions, have been proposed, but none has as yet been put forward that does not involve suffering for the elephants and disruption of the harmony with their environment in which they formerly lived. So it is hard to see a happy outcome for these creatures who tread so peacefully with trunks swaying, huge ears in gentle motion, plucking as they go leaves and grass and branches – great grey shapes, like tranquil battleships, moving up towards their sleeping quarters in the rocky ravines.

On our last evening in Manyara we drove along the lake shore, passing first a clutch of tourist buses encircling a pride of lions. Manyara lions are famous for lying draped about the branches of acacia trees, probably to keep cool and possibly to avoid being pestered by tsetse flies. (They have not fallen victim here to attacks by *Stomoxys*.)

We came to the sandy mouth of a stream where a great concourse of pelicans and cormorants was gathered. The pelicans had spread out in the shallow water of the lake in tens of thousands. They surged into the air in great sweeping clouds tinged with pink, making as they took off a sound like booted armies treading the ground, against a background of dark blue storm-clouds and distant misty blue hills. These huge birds look clumsy and rather ugly on the ground, but fly most gracefully.

Cormorants in thousands lined the muddy river bank as if waiting for a supper-bell to summon them. Their eyes were green. Their white throats pulsated like bellows, in and out – a cooling mechanism? Marabou storks, who stand five feet tall, stalked to and fro amongst them, and amongst

Overleaf: Elephant bulls eventually reach a height of thirteen feet and weigh six tons.

white and black wood ibises, white-all-over spoonbills gobbling with their spatulate bills, groups of gulls, and flocks of both spur-winged and Egyptian geese. The air was noisy with their mingled sounds – chatterings, wing-beats, landings and takings-off, air-beatings, honkings, splashings. Such a profusion of bird-life is overwhelming. I thought sadly of my English bird-table and its three or four pairs of tits, several sparrows, a few greenfinches, a chaffinch, a single robin. The appearance last winter of a solitary greater spotted woodpecker was an event, and the subject of telephone conversations with a neighbour whose bird-table competed with mine for his favours. How many birds were feeding on or near this lake-shore and in the lake? Half a million?

A hippo grunted at the river's mouth, slowly making his way upstream. We made our own way southwards along the shore among black stumps of trees killed by floods, to say goodbye to the elephants. In the far distance, where cliffs on our right stooped nearest to the lake to leave a narrow strand of grassland, Hugo had spotted a party of them on a small promontory, mustered for their evening feed before retreating to the wooded escarpment.

Several families had gathered to drink from a hole they had scooped out with their trunks from the bed of a stream. Having quenched their thirst, they plucked tendrils of a short, creeping grass with their trunks, and thrust them into their mouths. The calves were doing likewise but with much less concentration, breaking off to play and pester their elders. A giraffe had joined the party; she had a tiny toy-like calf that could not have been more than a few days old. Already it was showing signs of independence and quite clearly wanted to play with the baby elephants. It kept approaching them and being retrieved by its mother. The baby elephants didn't respond.

As usual, the elephants paid no attention to our Land-Rover and we watched them until the sun dropped behind the cliffs, their shadows vanished and the word was given – no doubt by the senior matriarch – to seek the night-time safety of the dark ravines. In their orderly fashion they formed into a file, two or three abreast, calves by their mothers' flanks, older children in her footsteps or not far from them, and with their deliberate gait passed within ten or fifteen yards of the Land-Rover, never once turning a head or stretching an inquisitive trunk to explore. Their manners were impeccable. Well fed, and watered, their mood was relaxed.

They passed so close that we could see each eyelash clearly, and the wrinkles on their heads, the scars and old slits in their ears and the blood-vessels under the skin, like the veins of a leaf, by which each

Wood ibis, pelicans and cormorants nest by their thousands in Lake Manyara National Park – famous for its enormous variety of birds and mammals.

individual elephant can be identified. Despite their indifference to our presence one was aware of their awareness, and knew that any sudden move or noise on our part would cause the whole group, numbering thirty or forty, to raise their trunks, break into a loping trot and be away and gone into the bush in the flick of a tail.

So goodbye, Manyara elephants. Another day is over, those big bellies filled with tasty fruit, with bark and branches, grass and seedpods, herbs and leaves, such a variety to choose from and to relish; mud-baths have been taken, children suckled, a drowsy siesta in the shade enjoyed, sunrise to sunset passed without alarums and excursions. May it so continue until the little calf we saw getting in the way of her elders, rolling under their bellies with her legs in the air, grows to be a matriarch with her grandchildren and great grand-children around her. May she live to lay her bones in peace when old age comes. May hunters with guns and poisoned arrows, may accident and mishap, may starvation as the woodlands shrink and wither, pass her by. May the hungry generations pressing in upon her sanctuary never tread her down.

# Eden under Siege?

What does the future hold for all these animals? And for their sanctuaries, the national parks? Can both survive?

Books, newspapers, journals, scientific papers: discussions, conferences, declarations – millions and millions of words about these subjects have been poured forth and the river has not yet run dry. Conservation has become one of the great issues of our time. Fifty years ago the word was little used, and as likely to be applied to crystallised fruit as to the saving from destruction of everything from the habitat of a rare orchid or a Victorian cinema to the forests of the Amazon basin or the elephants of Africa. We are all conservationists now – in theory. In practice, and with a few exceptions, battle after battle to preserve one or other of the earth's natural resources has been lost, or at best has led to compromise or a postponed decision. Year after year, the threat intensifies; unless we can contain it, conservationists must always be an army in retreat, defending positions until these are no longer tenable and then falling back to another line of defence.

To avoid extinction, prey animals evolved so that most would escape their predators. The development by man of destructive weapons has led to the wholesale extinction of many species with most others being threatened.

Crocodiles used to be common throughout Africa but have now been exterminated in most areas. Some of the largest, up to twenty feet long, still occur in the western Serengeti.

For many centuries, indeed millennia, the population of the world grew slowly, as the human species learned gradually to master and then exploit the environment and to deal with such natural checks as famine and disease. The rise accelerated and then, suddenly, came an explosion. The world's human population doubled in thirty-six years, rising from approximately two to four thousand million between 1940 and 1976. Barring some major disaster such as nuclear war, by the year 2000, not very far away, it will have reached six thousand million, and will go on rising ever faster year by year. Before this onslaught of people the forests of the world are going down, rivers are shrinking, minerals diminishing; the wilderness is in retreat. To feed these ever-multiplying millions, more and more land must be pressed into service. Techniques must be developed to make use of poorer land, wetlands must be drained, plains fenced into paddocks for domestic livestock, rivers dammed, and so on. Everywhere, humans are invading the last remaining habitats of the wild animals. And when the habitat goes, the wild animals go too.

However, there are national parks, supposedly inviolate. But are they? I have already mentioned poaching, a serious threat and one that won't go away. As animals with saleable tusks, horns or pelts grow scarcer, so do the rewards grow greater, and poachers become bolder and better equipped. So, also does temptation thrive to turn a blind eye or, more lucrative, to get in on the racket of trading in illegal trophies for which there is a world demand. Everyone from lowly paid game scouts in the parks to government ministers in their grand offices is exposed to this temptation – and many have succumbed.

But an even greater threat than poachers to the survival of the wild animals is the human womb. All these mounting millions need land for crops and pastures for livestock. One vital foodstuff, protein, can now be made synthetically from vegetable matter and from dried bacteria. So, in theory, we could do without meat. But that day is a long way off, and non-protein foods must still be grown. Africans need land, more and more of it as populations swell. Ever since they moved out of the hunter-gatherer stage of human development, they have been peasant farmers cultivating their *shambas*, and nomadic pastoralists living off their flocks and herds. And so do the great majority remain.

The best defence for most of the African parks lies in the poverty of their soil and the harshness of their climate. Many, though not all, were established in regions that were unoccupied or lightly populated because the soil was too poor for subsistence agriculture, rainfall too low, water too scarce, or because tsetse flies carrying diseases lethal to cattle had kept out the pastoral tribes. Humans cannot change climate – except for the worse, by destroying forests – but they can improve soil fertility, breed plants adapted to harsh conditions, immunise cattle against tsetse-borne disease and, with mechanisation, grow crops in regions formerly written off as wastelands. All this costs a lot of money and, for that reason, has largely remained a possibility on paper rather than a reality on the ground. But there are beginnings. There are now wheat-growing and range management projects in Maasailand and, as pressure mounts, cultivation in one form or another will continue to spread from more fertile regions into the less fertile ones. This seems inevitable. 'A man must live'. Few are likely to echo the words of the Compte D'Argenson who replied: '*Je n'en vois pas la nécessité.*'

Tourism is, of course, a powerful ally of the parks. In East Africa it has become a major industry, almost if not quite at the top of the league, earning the foreign exchange which is so badly needed. If all those

A serval cat in the Ngorongoro Crater during the short 'spring' rainy season.

The zebra migration at the start of the rainy season in the Serengeti.

zebra-striped buses and hired Land-Rovers ceased to quarter the parks in search of animals for their occupants to see and photograph, most of the tourist industry would fade away. The governments know this and so will, presumably, continue to preserve and develop their parks – for as long as they are able. This proviso must be added because the people they govern want land and grazing, not tourists and foreign exchange, and do not always grasp the need for the latter.

Meanwhile, the parks themselves have serious internal problems. I have already mentioned one: the paradox that some species whose overall numbers are declining, are multiplying in the parks to such an extent that they threaten the survival of their own habitat. Every farmer knows that if he crowds too many cattle on to a pasture, that pasture will get trampled, eaten out and eventually destroyed. So he moves them on well before that point has been reached. But animals in parks can't move on. They must stay within their limits and, so long as poaching is controlled, they multiply. If they move out of the parks they quickly get killed.

In former times, before human fecundity got out of hand, one of the ways in which many species of animal kept a balance between their food supply and their appetites was by migration. Early hunters and explorers in South Africa likened the migration of the springbok to a brown flood rolling across the veld – it was said that up to five million springboks might have been on the move. All gone now. Elephants migrated on a scale to match their bulk, also covering enormous distances. I have childhood memories of a glimpse of hundreds of elephants, I have no idea how many, moving without apparent haste and yet with remarkable speed along a regularly used migration route that ran between the forests of the Aberdare mountains and those of Mount Kenya. That sight will not be seen again.

The migration of the wildebeest is the largest of the few migrations that do survive, which is why it is so important to preserve it. At present, the scale looks so great that nothing could threaten it, and perhaps nothing will. One snag is that the migrating wildebeests do not stay within the parks – two are involved, the Serengeti and the Maasai-Mara in Kenya. The animals' terrestrial unit is the ecosystem, and less than half the Serengeti ecosystem lies within the National Park. Their migration takes them into areas of human settlement, especially in the west towards Lake Victoria, and in the south-west in the Maswa Game Reserve. (There are important differences between game reserves and national parks. In the former, hunting under licence is generally permitted and humans are not excluded; a recent report published in the journal *Oryx* stated that the Maswa reserve is 'seriously threatened by invading illegal settlement with a fast-growing population cultivating land and felling trees'. The report adds that in the Ngorongoro Conservation Area 'the Maasai have taken to poaching, both for subsistence meat and for trophies to sell – skins, ivory and rhino horn. In both places the guards are so poorly equipped they can do little to stop poaching.' The Maasai have, in fact, been poaching rhinos for some time.)

Another potential threat lies in the fact that the Maasai do not love the wildebeest. They consider that these animals are in possession of vast areas of grazing that would be much more usefully employed supporting unrestricted herds of Maasai cattle. Also they know that wildebeests carry the infection of a malignant catarrh to which their cattle are highly susceptible. So far as the Maasai are concerned the only good wildebeest is a dead one, and that also goes for almost every other species of herbivore. If the Maasai did not kill them in the past, that was because there was much less competition between wild and domestic animals and room enough for all, and they did not favour wild animals' meat.

At present the bitter complaints of the Maasai are mainly disregarded because the maintenance of parks and wildlife is a key part of government policy. Tanzania's President, Mwalimu Julius Nyerere, is a whole-hearted conservationist and, since Independence in 1961, his government has given consistent support to the policy of preservation, hampered only by an equally consistent shortage of cash. Despite these good intentions, there is a fragility about the situation disturbing to lovers of parks. Can Maasai discontent be contained forever? Or the Maasai persuaded to change their minds? Only the very optimistic would feel sure of a positive answer.

In some of the East African parks, notably the Kabalega Falls National Park in Uganda and the Tsavo National Park in Kenya, overcrowding in the 1960s became so acute that the animals appeared to be turning their habitats into deserts. Elephants were killing the trees by tearing off their bark or knocking them down, and degrading forest and acacia woodland into open savanna, thus altering entirely, some thought irretrievably, the nature of the country and also the elephants' diet. In Tsavo, which has a much lower rainfall than in Kabalega, elephants had all but stripped the country bare of bush and trees. Then came one of Africa's cyclical droughts, a severe one, and thousands of elephants and hundreds of rhinos died of thirst and starvation. The park presented a tragic scene of barren plain, dead trees, dried-up river-beds and skeletons. A similar situation arose in regard to hippos in the former Queen Elizabeth, now

The wildebeest migration of about two million animals gives some idea of what the bison herds must have looked like in North America before man all but exterminated them.

Ruwenzori, National Park in Uganda. The habit of hippos is to spend the day wallowing in shallow muddy water, and to come out at night to graze on the surrounding grasslands. The hippo population had built up to a point where their feeding areas had been eaten and trampled to death and were bare of vegetation.

The farmers' remedy for overstocking is culling – reducing the number of livestock on his land. Could the same principle be applied to elephants and hippos in national parks?

No issue has aroused more controversy among wildlife experts and the interested public in recent years than that of 'cropping', or killing, animals in national parks that appear to be destroying their habitat. There are strong arguments on both sides. The pro-cullers point out that to kill the animals quickly and cleanly with a skilled shot is more merciful than allowing them to starve to death. It is also less wasteful. Most Africans are short of protein; many of them relish elephant meat. So if the animals can be butchered properly and the meat taken quickly to the nearest market, the people will enjoy the protein, the croppers will make their profit (for this is a commercially run operation), and in the long run culling will benefit the elephants whose numbers will be stabilised and whose habitat will survive.

Culling of wild animals is not a new idea; it has been practised for years in the Kruger National Park in South Africa; in the Wankie Park in Zimbabwe and, intermittently, in several other southern African parks, as well as in national parks in other countries such as the United States. The African frontiers of the policy of culling and not-culling appear to have been drawn along the Zambezi river. In southern Africa culling remains, in general, an accepted practice; in eastern Africa opinion has, in the last decade or so, swung away from a belief in shooting overcrowded animals in national parks and towards a policy of letting nature take its course. Nowhere in eastern Africa is shooting in the parks a part of present wildlife policy.

The shooting policy has obvious disadvantages. However skilfully carried out, there is a danger that a wounded elephant might escape the *battue* and take his revenge on a tourist bus. Fear of man, provided that he stays in a vehicle, has to a considerable extent been replaced by trust, a trust that might well be shattered if man resumed the role of hunter that he has hitherto foregone in parks. If the animals took to their heels when a vehicle appeared, the tourists, metaphorically speaking, would soon take to theirs. The parks would be degraded, in the opinion of the anti-cullers, from wildlife sanctuaries to little more than game farms run by commercial concerns hoping to sell meat at a profit. The credibility of the whole park idea would be destroyed.

Pro-cullers admit these dangers, but ask what happens when the habitat is destroyed and the elephants die? Parks are run by trustees, with the animals and plants as wards; when the future of the wards is endangered, it is the duty of a trustee to intervene. Only, reply some at least of the trustees, if they are reasonably certain that the intervention will succeed. In our present state of ecological knowledge they cannot be certain; some even think that intervention may, in the long run, do more harm than good. There is nothing static about the processes of nature.

Topi often stand on top of termite mounds to spot more easily an approaching predator.

'National parks are not pictures on a wall,' it has been said. 'They are dynamic biological complexes with self-generating changes.' Plants and animals, soil and climate, all are in a constant state of modification and adjustment. The heavy hand of man may reverse or distort the adjustments to troubles for which nature may have its own remedies.

Africa is an ancient land, it has seen many cycles and changes, periods of drought and periods of flood. Lakes have come and gone; bush has given way to savanna and changed back into bush again, and always the animals and the vegetation have adapted and, in the long run, survived. To take one moment in the long march of time – the present – and suppose that as things are, so they will remain, is a false assumption. History is full of bright ideas that went wrong, such as the introduction of rabbits into Australia, of dogs and pigs into the Galapagos and South Sea islands, and so on. If, years later, you find that the bright idea was a blunder, the harm has been done.

And then, there is the hazard of the unexpected. Populations crash. Sometimes the reason is obscure, sometimes plain for all to see. Not much more than a decade after the successful cropping of elephants and hippos in Uganda's parks the government, if it can be called that, of Idi Amin collapsed, the Tanzanian army invaded, law and order broke down, chaos ensued and most of the park animals were destroyed. The elephants were reduced to a few small, frightened groups huddling round park lodges – not decimated, which properly means reduced by one tenth, but reduced by ninety per cent. In the Kabalega National Park an estimated nine thousand elephants were reduced to a mere one hundred and sixty. About a decade after elephant cropping had been proposed in the Tsavo National Park, drought drastically reduced the number of elephants and many other species in that park. Death from thirst and starvation is certainly more cruel than an accurate brain-shot, but nature is cruel, a fact accepted by those who believe in letting nature take its course in national parks.

So elephant populations crashed in Uganda and in Tsavo. Then the Tanzanian soldiers left Uganda and, though it could not be said that law and order were fully restored, the surviving elephants were allowed respite. The drought in Kenya ended in a bumper rainy season, Tsavo National Park turned green once more, rivers flowed and vegetation started to regenerate. Within a decade, elephants and other animals, except for the unfortunate rhinos, began to build up their numbers. Once again, the cycle was turning.

The elephant population crash in Uganda was an example of the hazard of political unrest, a fact that has by no means been eliminated in Africa (or indeed anywhere else). Then there is the factor of epidemics. In the late

178

All members in an elephant family unit are constantly aware of any difficulties a calf may get into and here, the mother assists her offspring to climb up a steep bank.

1880s rinderpest swept through much of southern and eastern Africa and millions of ungulates succumbed. While this did not affect the elephants, it reduced the buffalos, wildebeests and many other species, as well as the cattle, to little more than vestigial populations. These built up over the years but, in the opinion of some, may not have reached their pre-rinderpest plenitude, and now the disease has returned. Sweeping down from the Sudan it has affected buffalo and eland but, so far, not wildebeest. To go back somewhat further, to the Pleistocene age, it has been suggested that these African bushlands and savannas probably supported a greater concentration of animals than they support today; in fact Dr John Owen has suggested that we may be seeing, in such fragments of relatively

A kori bustard catches insects fleeing from a destructive bush fire started by man.

man-free Africa as remain, a return towards what he has called the 'Pleistocene abundance' of animal life.

Fragments – that is the trouble. National parks, even large ones like Serengeti, Tsavo and the Parc National Albert, are indeed but fragments of the African continent, and some are mere specks. Their enlargement is an urgent need, but is now very nearly impossible; in fact all the pressure is coming from the other direction, to excise not enlarge. In the words of Dr Richard Laws: 'The distribution of man and elephant populations has changed from one characterised by human islands in a sea of elephants to increasingly small islands of elephants in a sea of people.'

Some of the wisest and most experienced ecologists believe that there is an alternative to culling, an alternative produced by the elephants themselves – a 'self-regulatory mechanism' as it is called. When trees and woodlands destroyed by elephants in the Kabalega Park were replaced by other vegetation, mainly grasses, these became the staple of the elephants' diet. Then several thousand of the animals were shot; before the butchers did their work the scientists extracted various organs, analysed them, and found that significant changes were taking place. The effect of these was to delay the maturity of females and to lengthen the periods between calvings – a form of birth control. This was an adjustment to changes in the habitat. Regeneration of bush and woodland in Africa can be very rapid. Carried on over a much longer period, this self-regulatory mechanism might have led eventually to a return of the trees. The culling of the elephants, it was suggested by ecologists, may have checked the animals' natural response to change and in the long run made matters worse.

Nevertheless, parks must be managed to a greater or a lesser degree. It is impossible just to step back and do nothing, however desirable this might be. There are the tourists, for one thing, poachers for another. And there are other matters, such as fire. Anyone who has seen a grass or forest fire sweep over the plains of Africa must fear its ferocity and pity the countless creatures who are immolated. Fire leaves behind it smouldering tree-trunks and blackened plains; then come storks and cranes and other birds to gobble up charred and maimed insects. It is a pitiless sight. Surely control of fire must be essential if we are to fulfil the purposes of national parks?

Control is indeed essential: control but not elimination. Fires have been going through African bush and veld since time immemorial and it is one of the elements that have shaped the ecosystem. Fires help to keep a balance between thick bush and grassland; they destroy parasites, they provide fresh, succulent pasturage when (and how quickly) grass springs up through the blackened stubble. They may even provide for the regeneration of trees they have destroyed. There are varieties of tree whose seed-pods must experience fire before they can germinate. Vegetation has developed very cunning methods of survival. There is, for example, a variety of acacia that has developed a thick callus just below ground from which fresh roots emerge when fire has destroyed the tree; and there is a coarse, fibrous variety of grass whose thick tussocks act as nurseries to the seeds of other plants which, when the fire has passed and rain has come, germinate and start to re-colonise the scorched earth.

If fires in the past in Africa have contributed towards the balance of nature, in the present they show signs of getting out of hand. In those parks that lie in regions of low rainfall they have increased so much in recent years, owing to the increase of humans on the perimeters, to poachers and to honey-hunters (who light fires to smoke bees out of trees) and to other causes, that they threaten habitats which formerly survived less frequent burnings. I was shown the stump of a tree on the Serengeti that had lived, according to its rings, for thirty-seven years – and grown no more than a foot high. Every time it had started to put on growth again another fire had come along and scorched it.

Management of parks is a relatively new art-cum-science and there is still a lot to learn. Mistakes have been made, and no doubt others will be made. When an especially severe drought laid waste much of the Tsavo National Park in the early 1960s, when most of the rivers had dried up and animals were dying in droves, funds were raised to install watering-points fed by boreholes to enable thirsty and hungry animals to spread out into areas far from natural rivers and water-holes where some food was still available. Many lives were saved. Twenty years later, many ecologists have come to believe that, in the long term, this humanitarian act made the animals more vulnerable, not less, to the rigours of their environment. Droughts are one of nature's checks on population growth. More surviving animals will mean more hungry mouths to feed when the climatic cycle brings round the next drought; therefore greater suffering and more skeletons.

Drought and floods, fires and overcrowding, elephants, poachers, epidemics and all the other problems that African parks must continue to contend with are likely to count for less in the long run than politics. The parks were set up by law, and a simple law could at any time abolish them. Nothing seems less likely, we can thankfully say, at the present time. Yet – 'At my back I always hear not time's chariot hurrying near but the relentless lapping of a rising tide of people wanting land, wanting meat, wanting timber, wanting settlements and not wanting antelopes and lions. The survival of the parks depends on the will of governments to maintain a sea-wall strong enough to hold back the lapping, perhaps one day pounding, waves, and of acceptance by peasant farmers of the need to preserve, intact, areas where wildlife can thrive. Some observers of the scene are gloomy. An American authority, Dr Thomas Ofcansky, wound up his study of the history of wildlife preservation in East Africa with the words: 'It can only be concluded that man and animal cannot live side by side in peace and harmony. The only realistic wildlife preservation policy is therefore one that would delay rather than prevent the destruction of East Africa's fauna.'

We must hope that this conclusion is over-gloomy. The last century has seen a steady growth in the number of areas where wildlife is protected, in the number of countries which have set aside protected areas, and in the enthusiasm of the governments concerned for the concept of protection. The world's first national park, Yellowstone, was proclaimed in the United States in 1872, followed by the Banff park in Canada thirteen years later. It was not until 1926 that a former game reserve in South Africa became the Kruger National Park, the first in Africa; and not until after the Second World War that the idea really caught on all over the world. A positive rash of parks spread across the continents until, by 1982, the overall figure of 'protected areas' had reached the figure of about one and a half million square miles, or over two per cent of the terrestrial area of the globe. Taking game reserves and national parks together, plants and wild animals enjoy some measure of protection, theoretically at least, in about eleven per cent of Tanzania's total area. Zimbabwe's comparable figure is twelve per cent.

One of the most recent countries to expand its parks is Indonesia, which between 1972 and 1982 added eleven new ones to its existing five to make a total of nearly nineteen and a half thousand square miles under protection, or six per cent of the total land area. At the third World National Parks Congress held in 1982 in Bali, delegates from sixty-eight countries agreed a stirring declaration that called on all nations of the world to strengthen and expand their parks, and to confer upon them a truly lasting security. Of course it is one thing to make stirring declarations and draw lines on maps, and another to resist the mounting pressures I have mentioned and translate declarations into realities. Nevertheless, parks have become a source or pride to the countries in which they lie, and something of a status symbol among nations. There are still 'protected

Overleaf: Are these really the last days in Eden? We fervently hope not.

183

areas' where forests are being felled, habitats destroyed and where human settlements are creeping in. Yet if there is such a thing as world opinion, it condemns these lapses. The Tasmanian government's proposal to build a dam and install a hydro-electric plant in one of the few remaining large wilderness areas in Tasmania aroused an outcry in the world at large. When earth-moving and tree-felling machinery moved in, groups of protesting local conservationists were joined by Britain's Dr David Bellamy who flew to Tasmania in a blaze of publicity, got himself arrested and succeeded in getting the project temporarily halted. Then came a change of government in Australia and an announcement that the scheme was not to go ahead. This may be only a reprieve. Nevertheless, twenty years ago, Tasmania would just have built its dam without any fuss or outside interference. Similar public protests and scientific objections have recently succeeded in getting the Malaysian government actually to abandon its intention to build a dam in peninsular Malaysia's only National Park, Taman Negara, which would have flooded one third of the area and so destroyed both virgin forest and the wildlife dwelling there. Conservationists can only hope that the threat will not be renewed. But how sacrosanct are parks when proposals to destroy large parts of them are advanced by the very government supposed to protect them forever; and only withdrawn at the eleventh hour and as a result of public pressure?

Nevertheless, attitudes are changing. 'Taming the wilderness' used to be considered a worthy, even noble, aim; today 'plundering the wilderness' has become something like a crime. Now there is a World Conservation Strategy whose aims, in the words of conservationist Max Nicholson, are to counter the 'false values that threaten human survival by their short-sighted materialism . . . by reintegrating mankind as part of the natural world'. A formidable aim indeed, in a society which seems to have headed rapidly in the opposite direction towards a most unnatural world of computers, nuclear weapons, robots, high technology, space probes and the like.

In bringing about this shift in world opinion the wildlife photographer has played, and will continue to play, a key part. He brings into our homes pictures so vivid, so accurate and beautiful, that few who see them can remain indifferent to the fate of the birds and beasts, plants and insects, fish and reptiles he portrays. Through his lenses we see into a world hitherto unknown to most of us, a world whose survival can be seen to be essential to the enjoyment and enrichment of mankind. How fortunate it is that the technical development of photography reached its present stage

186

of excellence just at the time when the need to make a record of diminishing habitats and endangered species became so important. Fifty years ago, lenses were crude and limited in scope, and wildlife abundant. Photographs then were trophies, more interesting and less cumbersome than the horns that used to adorn – or deface – the walls of trophy-hunters' mansions, but of little scientific or artistic interest. Today they are both artistic achievements and invaluable records of a vanishing world. Were the wildlife of our national parks to disappear tomorrow, we should still have these records of their incomparable variety to marvel at and enjoy.

So the wildlife photographers of today are much more than just entertainers. They are chroniclers of the natural world who preserve on film a record of its riches and diversity, perhaps destined in future to be as valuable in its way as are the chronicles of human achievement preserved on parchment and paper. They catch the moment, the scene, the incident that may never come again. As I have tried to show, the good wildlife photographer must possess not only the artist's eye but also something of the scientist's knowledge, patience and precision. Also he is, perhaps unconsciously, a powerful champion of the conservation cause.

The muster of professional wildlife photographers, though growing, is not large, and those in the top flight, as in any profession, are few. Among them Hugo van Lawick is certainly to be numbered. He, certainly, would not claim pre-eminence; there are others no less skilled and successful in the field. Most of such photographers are based outside the parks and the world is their oyster so far as their work is concerned. Hugo is exceptional in that he has specialised on a particular region and lived within it for much of his working life. Thus he has built up a body of experience and a collection of pictures which, by narrowing the range, achieve a greater depth of focus than those who use, as it were, a wider-angled lens.

In this commentary written around Hugo's pictures, I have tried to suggest how a wildlife photographer goes about his business, in this case a particular photographer in his chosen venue, the Serengeti plains and their surrounding ecosystem. Other commentaries could be written around the working lives of other photographers, each one of whom stamps upon his work his unique personality. Through the same lenses, each individual pair of eyes creates on film a different vision of a similar scene. Therein lies the art of the photographer.

The good photographer must be a dedicated man. If there is one word that sums up Hugo's quality in his profession, that is it – dedication. These pictures, selected from tens of thousands, are the proof.

# Bibliography

Brown, Leslie. *The Mystery of the Flamingos*. Country Life, 1959

Cole, Sonia. *Leakey's Luck*. Collins, 1975

Dawkins, Richard. *The Selfish Gene*. OUP, 1976; Paladin (paperback), 1978

Douglas-Hamilton, Iain & Oria. *Among the Elephants*. Collins & Harvill, 1976

Goodall, Jane. *In the Shadow of Man*. Collins, 1971

Huxley, Elspeth. *Forks and Hope*. Chatto & Windus, 1964

Huxley, Elspeth. *Whipsnade: Captive Breeding for Survival*. Collins, 1981

Jewel, Peter, & Holt, Sidney. *Problems in Management of Locally Abundant Wild Animals*. Academic Press, 1982

Kingdon, Jonathan. *East African Mammals: An Atlas of Evolution in Africa* (7 vols). Academic Press, 1971–82

Kruuk, Hans. 'A New View of the Hyena', *New Society*, 30 June 1966

Lawick, Hugo van. *Solo: The Story of an African Wild Dog*. Collins, 1973; Fontana (paperback), 1976

Lawick, Hugo van, & Goodall, Jane. *Innocent Killers*. Collins, 1970

Leakey, Mary. *Olduvai Gorge: My Search for Early Man*. Collins, 1979

Leakey, Richard. *The Making of Mankind*. Michael Joseph, 1981

Martin, Esmond, & Chryssee. *Run Rhino Run*. Chatto & Windus, 1982

Ofcansky, Thomas. *A History of Game Preservation in British East Africa, 1895–1963*. University of West Virginia (unpublished thesis)

Owen John. 'Some Thoughts on Management in National Parks', *Biological Conservation*, Vol. 4, Part 4, 1972

Russell, Walter. *Management Policy in the National Parks*. Tanzania National Parks, 1968

Saitoti, Tepilit Ole. *Masai*. Elm Tree, 1980

Schaller, George. *The Serengeti Lion*. Chicago, 1972

Serengeti Research Institute. *Annual Reports*, 1965–71

Tall Timbers Fire Ecology Conference, 1971. *No. 11: Fire in Africa*

Williams, John, & Arlott, Norman. *Field Guide to the Birds of East Africa*. Collins, 1980

# *Index*